Learning Node.js for Mobile Application Development

Make use of Node.js to develop of a simple yet scalable cross-platform mobile application

Stefan Buttigieg

Milorad Jevdjenic

[PACKT] open source *
PUBLISHING
community experience distilled

BIRMINGHAM - MUMBAI

Learning Node.js for Mobile Application Development

First published: October 2015

Production reference: 1231015

Published by Packt Publishing Ltd.
Livery Place
35 Livery Street
Birmingham B3 2PB, UK.

ISBN 978-1-78528-049-8

www.packtpub.com

Credits

Authors
Stefan Buttigieg
Milorad Jevdjenic

Reviewers
Danny Allen
Alex (Shurf) Frenkel
Siddique Hameed
Prasanna Ramanujam

Commissioning Editor
Nadeem N. Bagban

Acquisition Editor
Harsha Bharwani

Content Development Editor
Sumeet Sawant

Technical Editor
Tejaswita Karvir

Copy Editor
Vedangi Narvekar

Project Coordinator
Shweta H Birwatkar

Proofreader
Safis Editing

Indexer
Hemangini Bari

Graphics
Abhinash Sahu

Production Coordinator
Komal Ramchandani

Cover Work
Komal Ramchandani

About the Authors

Stefan Buttigieg is a medical doctor, mobile developer, and entrepreneur. He graduated as a doctor of medicine and surgery from the University of Malta. He is currently enrolled at the University of Sheffield, where he is pursuing a master's degree in health informatics. He has more than 5 years of experience working in international medical students' organizations, where he occupied various technical positions. He founded **MD Geeks**, an online community that brings together health professionals, developers, and entrepreneurs from around the world to share their passion for the intersection of healthcare and information technology. He is mostly interested in mobile development, especially for the Android and iOS platforms, open source healthcare projects, user interface design, mobile user experience, and project management.

Milorad Jevdjenic is a programmer, open source enthusiast, and entrepreneur. He studied computer science at the University of Gothenburg with a focus on formal verification methods. Currently, he works as a software developer in the medical sector and also does independent commercial and pro bono consulting. Milorad is passionate about technology. He looks upon open source, and open standards in particular, as the fundamental drivers that are needed to build better societies. When he is not on the computer tinkering with code, he enjoys hiking, sports, and fine whiskey.

About the Reviewers

Danny Allen is a full stack web developer with a focus on user experience design and implementation. He is a founder and director of the international consultancy, Wonderscore Ltd.

Skilled across a wide range of backend and frontend technologies, including Python and Django, JavaScript, AngularJS, Node.js, HTML5, and CSS3, his recent work has involved e-learning and government projects in the United Kingdom.

Danny currently lives and works in Barcelona, Spain.

His portfolio and contact details can be found at `http://dannya.com`.

Alex (Shurf) Frenkel has been working in the field of web application development since 1998 (the beginning of PHP 3.x). He has an extensive experience in system analysis and project management. Alex is a PHP 5.3 **Zend Certified Engineer** (**ZCE**) and is considered to be one of the most prominent LAMP developers in Israel. He is also a food blogger. You can view his blog by visiting `http://www.foodstuff.guru`.

In the past, Alex was the CTO of ReutNet, one of the leading Israeli web technology -based companies. He also worked as the CEO/CTO of OpenIview LTD, a company built around the innovative idea of breaching the IBM Mainframe business with PHP applications, and as a CTO and chief architect of a start-up named GBooking. He also provided expert consulting services to different companies regarding various aspects of web-related technology.

Frenkel-Online is a project-based company that works with a number of professional freelance consultants in Israel and abroad. Currently, their permanent staff comprises several professionals from Israel and abroad for the company's PHP projects, and a changing number of specialists in other programming languages for the rest of the projects.

FoodStuff.Guru is a pet project that brings not only high-style food, but also common food, to the web so that it can be reviewed by people for the people. The blog is multilingual and can be viewed by visiting `http://www.foodstuff.guru`.

Siddique Hameed is currently working as a full-stack engineer on Simplify Commerce (`http://simplify.com`), a payment gateway platform from MasterCard. In his diverse career experience, he has crafted software for Fortune 500 companies as well as startups with industry domains ranging from commerce, social media, telecom, bio-informatics, finance, publishing, insurance, and so on.

He is a passionate technologist who actively contributes to open source projects. He speaks frequently at tech events and meet-ups and mentors the participants of hackathons and code boot camps.

His current focus areas include AngularJS, Ionic, Node.js, HTML5, CSS3, Cloud computing, mobile applications, and the **Internet of Things (IoT)**. In his spare time, he likes to tinker with the Raspberry Pi and build DIY gadgets.

I dedicate this to my mom, dad, my beloved wife, Farzana, and my wonderful daughters, Fareeha and Sameeha!

Prasanna Ramanujam is a software engineer. He has a master's degree in software engineering. He is a full-stack developer, and he has been a Node.js developer since the release of Node.js version 0.2. He has helped architect and scale the Node.js application at companies in Silicon Valley. He has also published many private and public NPM modules.

He is passionate about building high-availability systems. He likes to work on asynchronous programming, distributed computing, and NoSQL databases. He also likes music, skiing, and water sports. He can be found on Twitter at @prasanna_sr.

I would like to thank my family members and friends for supporting me.

Also, my sincere thanks to Pooja Mhapsekar and the other members from Packt Publishing for giving me this opportunity as well as Shweta Birwatkar for coordinating with me throughout the journey.

---

www.PacktPub.com

Support files, eBooks, discount offers, and more

For support files and downloads related to your book, please visit www.PacktPub.com.

Did you know that Packt offers eBook versions of every book published, with PDF and ePub files available? You can upgrade to the eBook version at www.PacktPub.com and as a print book customer, you are entitled to a discount on the eBook copy. Get in touch with us at service@packtpub.com for more details.

At www.PacktPub.com, you can also read a collection of free technical articles, sign up for a range of free newsletters and receive exclusive discounts and offers on Packt books and eBooks.

https://www2.packtpub.com/books/subscription/packtlib

Do you need instant solutions to your IT questions? PacktLib is Packt's online digital book library. Here, you can search, access, and read Packt's entire library of books.

Why subscribe?

- Fully searchable across every book published by Packt
- Copy and paste, print, and bookmark content
- On demand and accessible via a web browser

Free access for Packt account holders

If you have an account with Packt at www.PacktPub.com, you can use this to access PacktLib today and view 9 entirely free books. Simply use your login credentials for immediate access.

Table of Contents

Preface

Node.js is a massively popular JavaScript library that lets you use JavaScript to easily program scalable network applications and web services. People approaching Node.js for the first time are often attracted by its efficiency, scalability, and the fact that it's based on JavaScript, the language of the Web. This means that developers can use the same language to write backend code. Also, it's increasingly being looked upon as a modern replacement for PHP in web development, which relies on fast-paced data exchange. This growing community and the large amount of available modules makes Node.js one of the most attractive development environments.

What this book covers

Chapter 1, *Setting Up Your Workspace*, explains how to set up your work environment to develop cross-platform applications by using the Ionic framework as the frontend tool, Node.js for the backend, and the integrated development environment, Atom.

Chapter 2, *Configuring Persistence with MongoDB*, goes through the necessary configurations that are needed to make an instance of MongoDB work with Node.js. You will learn how to set up security and a database, install the relevant MongoDB driver for Node.js, and communicate with the database from a Node.js instance.

Chapter 3, *Creating an API*, looks at how we can set up a uniform interface for sending and receiving data and basic functionality on the Node.js server by building an **API** (**Application Programming Interface**) that exposes it. We will cover the basic REST topics and show you how to configure routes to perform simple read/write operations on our data.

Chapter 4, *Securing Your Backend*, shows that this type of remedies can be achieved by building a basic security mechanism in order to control user access. Specifically, we will deal with token-based authentication and show you how this makes it easy to limit access to your backend. In doing so, we will introduce the concept of roles and how they figure in our authentication scheme.

Chapter 5, Real-Time Data and WebSockets, shows you how to enable real-time data communication using WebSockets. This will allow your server to directly communicate with connected clients without having to perform polling on the client side.

Chapter 6, Introducing Ionic, covers all the basic essentials that are needed to set up a working environment, which is required to efficiently create and share Ionic apps.

Chapter 7, Building User Interfaces, takes the template project that we created in the past few chapters and modifies it to something closer to what we envisioned by altering the appearance of the user interface. In addition to this, we will also start experimenting with the Ionic project code to deeply understand what the project is made up of.

Chapter 8, Making Our App Interactive, covers a lot of ground, going into the details of AngularJS and learning more about the interaction of the model, view, and controller. We will also see how to use the Cordova plugins and ngCordova in order to access native features. Finally, we will also have a look at how to create services and use them in order to serve data to the users.

Chapter 9, Accessing Native Phone Features, discusses how to directly work with the Google Maps API in order to render and work with maps.

Chapter 10, Working with APIs, explores how to access external APIs in order to send and retrieve data. You will learn how to encode/decode data to/from JSON in our app in order to provide a standard interface for processing.

Chapter 11, Working with Security, introduces the concept of security in mobile apps in the context of authenticating and authorizing local users. We will have a look at how to implement a common login feature, which contacts a remote server (run by Node.js, of course!) in order to verify that a given pair of user credentials is correct. We then use this information in order to grant the user access to the rest of the application.

Chapter 12, Working with Real-Time Data, covers how to set up WebSocket communication through the mobile app in order to subscribe to dynamic notifications from a server. We will elaborate on how this helps us develop truly dynamic applications, such as chat apps. This chapter will further introduce the concept of push notifications, which will allow our app to get updates from a server on a dynamic basis.

Chapter 13, *Building an Advanced Chat App*, expands this mobile app and makes it more advanced by adding the features of chat rooms and notifications. In doing so, we will demonstrate how the concept of namespacing works on socket.io, which is one of the most important aspects of this library.

Chapter 14, *Creating an E-Commerce Application Using the Ionic Framework*, brings together the knowledge that you have accumulated from this book and implements it in an easy-to-use Ionic framework that can be used in your very own projects.

What you need for this book

You'll need the following software:

- Android Studio
- Android Software Development Kit
- The Ionic framework
- MongoDB
- Atom
- XCode

Who this book is for

This book is intended for web developers of all levels of expertise who want to deep dive into cross-platform mobile application development without going through the pain of understanding the languages and native frameworks that form an integral part of developing for different mobile platforms.

This book is also for developers who want to capitalize on the Mobile First strategy and who are going to use JavaScript for their complete stack.

Conventions

In this book, you will find a number of styles of text that distinguish between different kinds of information. Here are some examples of these styles, and an explanation of their meaning.

Code words in text, database table names, folder names, filenames, file extensions, pathnames, dummy URLs, user input, and Twitter handles are shown as follows: "We can include other contexts through the use of the `include` directive."

A block of code is set as follows:

```
angular.module('supernav.controllers', [])
.controller('MapCtrl', function ($scope) {
  $scope.mapCreated = function (map) {
    $scope.map = map;
  };
});
```

When we wish to draw your attention to a particular part of a code block, the relevant lines or items are set in bold:

```
.state('app.scala', {
  url: '/scala',
  views: {
    'scala-view': {
      templateUrl: 'templates/app-chat.html',
      controller: 'ChatController',
      resolve: {
        chatRoom: function () {
          return 'scala';
        }
      }
    }
  }
});
```

Any command-line input or output is written as follows:

```
console.log('Hello World!');
```

New terms and **important words** are shown in bold. Words that you see on the screen, in menus or dialog boxes for example, appear in the text like this: "Conclude this process by clicking on **Create Column**, and we are done!".

> Warnings or important notes appear in a box like this.

> Tips and tricks appear like this.

Reader feedback

Feedback from our readers is always welcome. Let us know what you think about this book—what you liked or may have disliked. Reader feedback is important for us to develop titles that you really get the most out of.

To send us general feedback, simply send an e-mail to feedback@packtpub.com, and mention the book title via the subject of your message.

If there is a topic that you have expertise in and you are interested in either writing or contributing to a book, see our author guide on www.packtpub.com/authors.

Customer support

Now that you are the proud owner of a Packt book, we have a number of things to help you to get the most from your purchase.

Downloading the example code

You can download the example code files for all Packt books you have purchased from your account at http://www.packtpub.com. If you purchased this book elsewhere, you can visit http://www.packtpub.com/support and register to have the files e-mailed directly to you.

Downloading the color images of this book

We also provide you a PDF file that has color images of the screenshots/diagrams used in this book. The color images will help you better understand the changes in the output. You can download this file from: http://www.packtpub.com/sites/default/files/downloads/1453OT_ColorImages.pdf.

Errata

Although we have taken every care to ensure the accuracy of our content, mistakes do happen. If you find a mistake in one of our books—maybe a mistake in the text or the code—we would be grateful if you would report this to us. By doing so, you can save other readers from frustration and help us improve subsequent versions of this book. If you find any errata, please report them by visiting http://www.packtpub.com/submit-errata, selecting your book, clicking on the **erratasubmissionform** link, and entering the details of your errata. Once your errata are verified, your submission will be accepted and the errata will be uploaded on our website, or added to any list of existing errata, under the Errata section of that title. Any existing errata can be viewed by selecting your title from http://www.packtpub.com/support.

Piracy

Piracy of copyright material on the Internet is an ongoing problem across all media. At Packt, we take the protection of our copyright and licenses very seriously. If you come across any illegal copies of our works, in any form, on the Internet, please provide us with the location address or website name immediately so that we can pursue a remedy.

Please contact us at `copyright@packtpub.com` with a link to the suspected pirated material.

We appreciate your help in protecting our authors, and our ability to bring you valuable content.

Questions

You can contact us at `questions@packtpub.com` if you are having a problem with any aspect of the book, and we will do our best to address it.

1
Setting Up Your Workspace

The overarching goal of this book is to give you the tools and know-how needed to efficiently construct modern, cross-platform solutions for your users. In this chapter, we will focus on the tools themselves, showing you how to bootstrap your development environment to tackle the challenges that we have in store for you in the remainder of the book. We will also give you a brief introduction to each tool in order to give you an idea about why they fit into your toolchain. If any such detail seems unclear at this point, do not worry. We will delve into everything you need to know about each tool in the relevant parts of the book.

After reading this chapter, you will know how to install, configure, and use the fundamental software components that we will use throughout this book. You will also have a good understanding of why these tools are appropriate for the development of modern apps.

The Node.js backend

Modern apps have several requirements, which cannot be provided by the app itself, such as central data storage, communication routing, and user management. In order to provide such services, apps rely on an external software component known as the backend. The backend will be executed on one or more remote servers, listen to network requests from the devices that run the app, and provide them with the services that requests require.

The backend that we will use in this book is Node.js, a powerful but strange beast in its category. Node.js, at the time of writing this book, is the only major backend that is written almost entirely in JavaScript, which in reality is a frontend scripting language. The creators of Node.js wanted a backend that could be integrated with the apps written in JavaScript as seamlessly as possible, and you cannot get much closer to that than Node.js. Beyond this, Node.js is known for being both reliable and high-performing.

In terms of architecture, Node.js is highly modularized and designed from the ground up to be extendable through plugins or packages. Node.js comes with its own package management system, **Node Package Manager** (**NPM**), through which you can easily install, remove, and manage packages for your project. You will see how to use NPM in order to install other necessary components later in this chapter.

Installing Node.js on different systems

Node.js is delivered as a set of JavaScript libraries, executing on a C/C++ runtime built around the Google V8 JavaScript Engine. The two come bundled together for most major **operating systems** (**OS**), and we will look at the specifics of installing it in the following sections.

> Google V8 JavaScript Engine is the same JavaScript engine that is used in the Chrome browser, which is built for speed and efficiency.

Windows

For Windows, there is a dedicated MSI wizard to install Node.js, which can be downloaded from the project's official website. To do so, go to the main page, navigate to **Downloads**, and then select **Windows Installer**. After it has downloaded, run the MSI wizard, follow the steps to select the installation options, and conclude the install. Keep in mind that you will need to restart your system in order to make the changes effective.

Linux

Most major Linux distributions provide convenient installs of Node.js through their own package management systems. However, it is important to keep in mind that for many of them, **Node Package Manager** (**NPM**) will not come bundled with the main Node.js package. Rather, it is provided as a separate package. We will show how to install both in the following section.

Ubuntu/Debian

Open a terminal and issue `sudo apt-get update` to make sure that you have the latest package listings. After this, issue `apt-get install nodejsnpm` in order to install both Node.js and NPM in one swoop.

Fedora/RHEL/CentOS

On Fedora 18 or later, open a terminal and issue `sudo yum install nodejsnpm`. The system will do the full setup for you.

If you are running RHEL or CentOS, you need to enable the optional EPEL repository. This can be done in conjunction with the install process, so that you do not need to do it again while upgrading the repository, by issuing the `sudo yum install nodejsnpm --enablerepo=epel` command.

Verifying your installation

Now that we have finished the install, let's do a sanity check and make sure that everything works as expected. To do so, we can use the Node.js shell, which is an interactive runtime environment for the execution of JavaScript code. To open it, first open a terminal, and then issue the following to it:

```
node
```

This will start the interpreter, which will appear as a shell, with the input line starting with the > sign. Once you are in it, type the following:

```
console.log("Hello world!");
```

Then press *Enter*. The **Hello world!** phrase should appear on the next line. Congratulations, your system is now set up for the running of Node.js!

Mac OS X

For OS X, you can find a ready-to-install PKG file by going to www.nodejs.org, navigating to **Downloads**, and selecting the **Mac OS X Installer** option. Otherwise, you can click on **Install**, and your package file will automatically be downloaded:

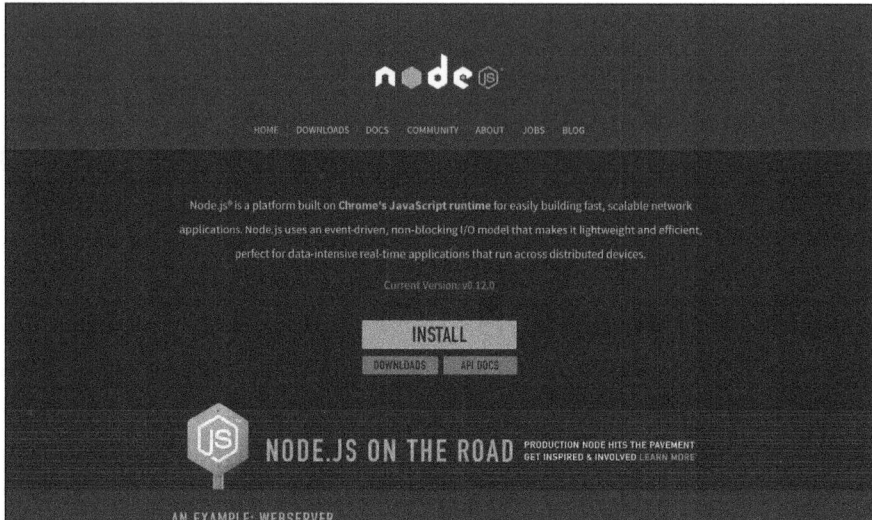

Once you have downloaded the file, run it and follow the instructions on the screen. It is recommended that you keep all the default settings offered unless there are compelling reasons for you to change something with regard to your specific machine.

Verifying your installation

After the install finishes, open a terminal and start the Node.js shell by issuing the following command:

```
node
```

This will start the interactive node shell, where you can execute JavaScript code. To make sure that everything works, try issuing the following command to the interpreter:

```
console.log("hello world!");
```

After pressing the *Enter* key, the **hello world!** phrase will appear on your screen. Congratulations, Node.js is all set up and good to go!

Setting up the Ionic framework and Cordova for Mac OS X

After installing Node.js on your Mac, proceed to open your command-line application and input the following command:

```
$ sudonpm install -g ionic
```

After inputting this command, you will be prompted to input your password as shown in the following screenshot:

If you have already set up the permissions for npm on OS X, you can install Ionic with the following command:

```
$ npm install -g ionic
```

The preceding command line should result in the following output:

```
● ● ●              ⬆ stefanbuttigieg — bash — 80×24
├── npm@2.1.3
├── tiny-lr-fork@0.0.5 (debug@0.7.4, faye-websocket@0.4.4, qs@0.5.6, noptify@0.0
.3)
├── prompt@0.2.12 (revalidator@0.1.8, pkginfo@0.3.0, read@1.0.5, winston@0.6.2,
utile@0.2.1)
├── connect@3.1.1 (utils-merge@1.0.0, parseurl@1.3.0, finalhandler@0.1.0, debug@
1.0.4)
├── request@2.51.0 (caseless@0.8.0, json-stringify-safe@5.0.0, forever-agent@0.5
.2, aws-sign2@0.5.0, stringstream@0.0.4, oauth-sign@0.5.0, tunnel-agent@0.4.0, n
ode-uuid@1.4.2, qs@2.3.3, mime-types@1.0.2, combined-stream@0.0.7, tough-cookie@
0.12.1, http-signature@0.10.1, form-data@0.2.0, bl@0.9.4, hawk@1.1.1)
├── serve-static@1.7.1 (utils-merge@1.0.0, escape-html@1.0.1, parseurl@1.3.0, se
nd@0.10.1)
├── xml2js@0.4.4 (sax@0.6.1, xmlbuilder@2.5.2)
├── unzip@0.1.9 (setimmediate@1.0.2, match-stream@0.0.2, pullstream@0.4.1, reada
ble-stream@1.0.33, binary@0.3.0, fstream@0.1.31)
├── vinyl-fs@0.3.7 (graceful-fs@3.0.5, lodash@2.4.1, mkdirp@0.5.0, strip-bom@1.0
.0, vinyl@0.4.6, through2@0.6.3, glob-stream@3.1.18, glob-watcher@0.0.6)
├── archiver@0.5.1 (lodash@2.4.1, readable-stream@1.1.13, lazystream@0.1.0, zip-
stream@0.1.4, file-utils@0.1.5)
└── gulp@3.8.8 (pretty-hrtime@0.2.2, interpret@0.3.10, deprecated@0.0.1, archy@0
.0.2, minimist@1.1.0, semver@3.0.1, tildify@1.0.0, chalk@0.5.1, orchestrator@0.3
.7, liftoff@0.12.1, gulp-util@3.0.3)
Stefs-Macbook-Pro:~ stefanbuttigieg$ ▮
```

Installing Cordova on OS X is very similar to installing Ionic. You can run the following command to install Cordova:

```
$ sudonpm install -g cordova
```

Setting up the Ionic framework and Cordova for Windows

Once you have installed Node.js, install Ionic on your Windows machine. The rest should be straightforward.

Open the command prompt and check whether you have npm installed by running the following command:

```
npm
```

Once you have ensured that you have successfully installed npm, you can go ahead and run the following command:

```
npm install -g ionic
```

---

[6]

This step should result in an output, which shows that you have successfully installed Ionic.

In order to install Cordova, you can also use `npm` and run the following command:

```
npm install -g cordova
```

Once you receive a successful output, you can go ahead and start setting up the platform dependencies.

> An experimental setup for Windows:
>
> In Windows, you will have the opportunity to set up a Vagrant package, which is a one-stop-shop for the installation of Ionic on your Windows machine. This is accessible at `https://github.com/driftyco/ionic-box`.

Setting up the platform dependencies

To set up the platform dependencies, you need to install Java, which is explained in the following section.

Installing Java

If you do not have Java installed or if your version is below 6.0, install the Java JDK by heading over to `http://j.mp/javadevkit-download`, a customized and shortened link, and choosing the version that applies to you.

The main recommendation for these projects is that you install a version of JDK 6.0 or higher.

Select the JDK for your OS. On an Intel-based Mac, you can use the following useful table to check whether your Mac is a 32- or 64-bit OS.

You can check for **Processor Name** by clicking on the Apple logo in the top-left corner of your screen, followed by **About my Mac**:

Processor Name	32- or 64-bit
Intel Core Solo	32 bit
Intel Core Duo	32 bit
Intel Core 2 Duo	64 bit
Intel Quad-Core Xeon	64 bit
Dual-Core Intel Xeon	64 bit

Processor Name	32- or 64-bit
Quad-Core Intel Xeon	64 bit
Core i3	64 bit
Core i5	64 bit
Core i7	64 bit

In the case of Windows, if you have a machine that was purchased in the last few years, you should go for the x64 (64-bit) version.

Setting up Android Studio for Android, Mac, and Windows

To set up Android Studio for Android, Mac, and Windows, follow these steps:

1. Go to the Android Developers site by visiting `http://developer.android.com`.

2. Click on **Android Studio**, where you will be directed to the landing page. Your operating system's version will be detected automatically:

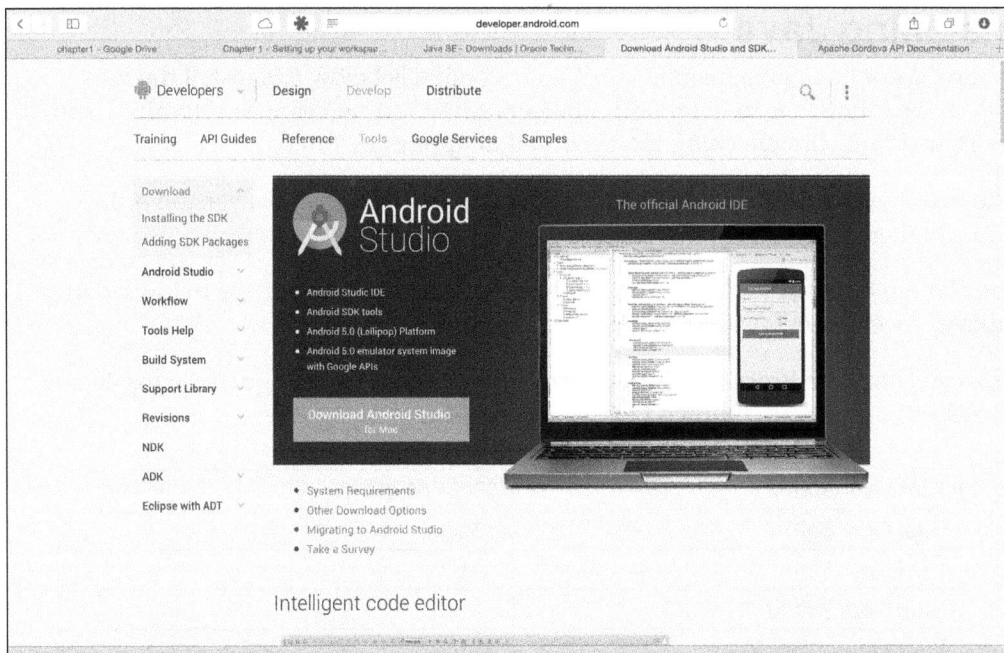

3. Accept the terms and conditions of the Software Use Agreement and click on **Download**:

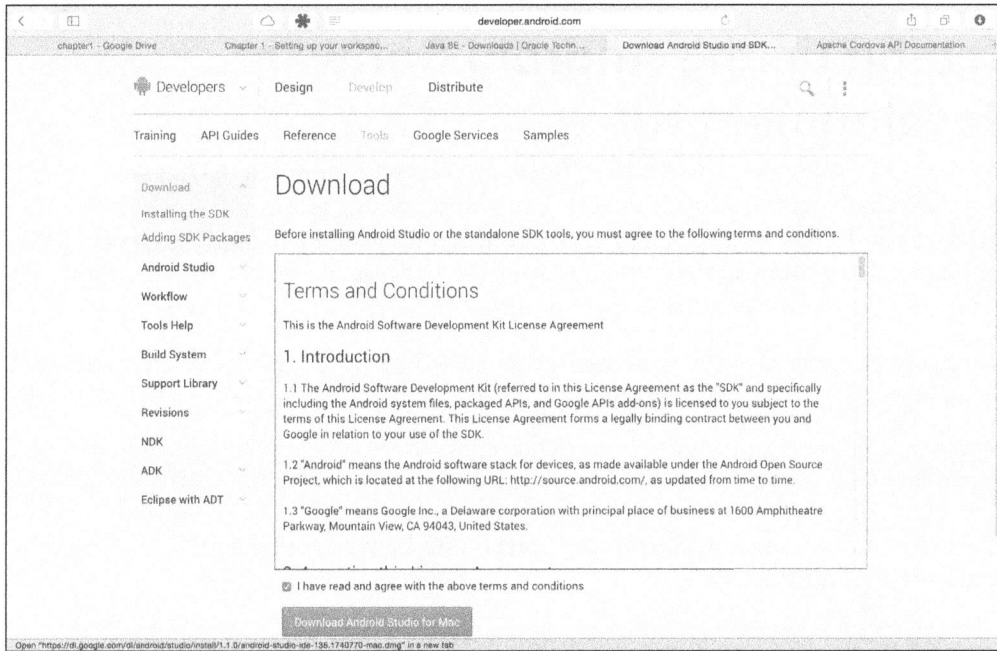

4. For Mac, double-click on the downloaded file, follow the prompts, and then drag the Android Studio icon into your **Applications** folder:

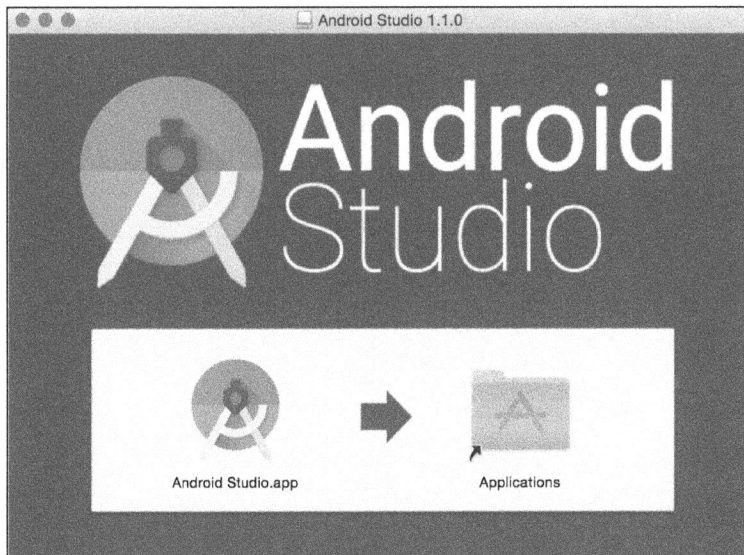

5. For Windows, open the downloaded file, and then go through the **Android Studio Setup Wizard** to complete the install.

Setting up the Android Software Development Kit

The process of setting up the Android **Software Development Kit (SDK)** has improved vastly with the introduction of Android Studio, as a number of software packages come pre-installed with the Android Studio install package. As a part of the preparation for getting started with our Android projects, it will be very helpful to understand how one can install (or even uninstall) SDKs within Android Studio.

There are a number of ways of accessing the SDK Manager. This can be done from the main toolbar of Android Studio:

Otherwise, it can be accessed from the **Start** menu by navigating to **Configure – SDK Manager**:

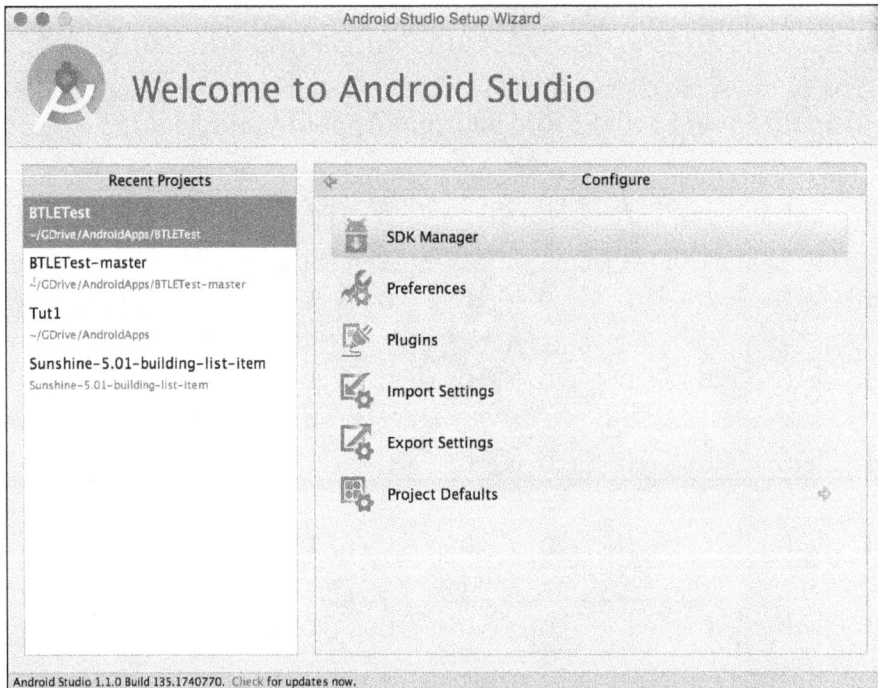

This is what the SDK Manager looks like. If you need to install a package, you need to check the mark of that particular package, click on **Install packages**, and then finally accept the licenses:

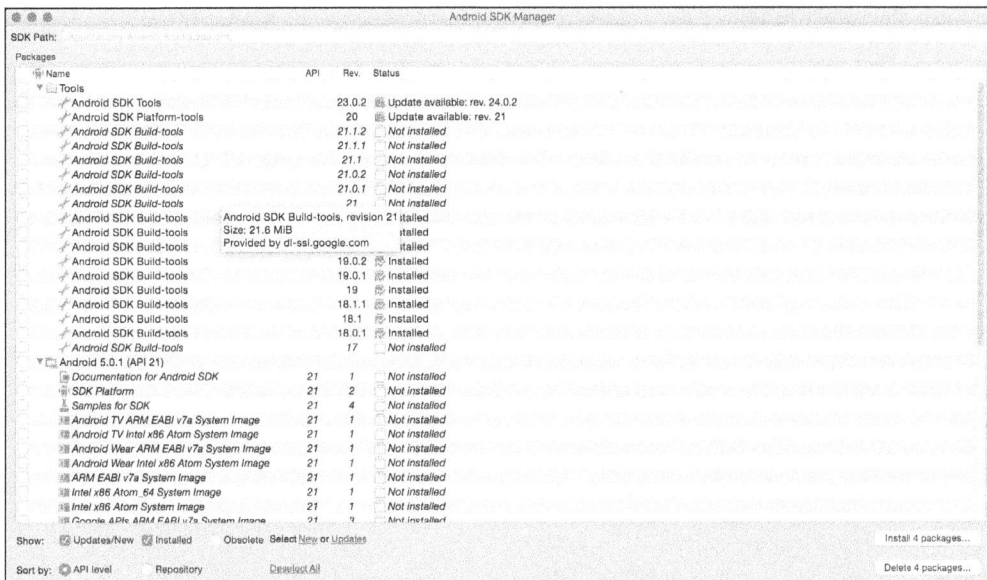

Setting up your physical Android device for development

The following are the three main steps that need to be taken in order to enable your Android Device for development:

1. Enable **Developer options** on your specific Android device

2. Enable **USB debugging**

3. Provide your computer with the necessary trust credentials with the installed IDE via secure USB debugging. (devices with Android 4.4.2)

Enabling Developer options

Depending on your device, this might vary slightly, but as from Android 4.2 and higher, the **Developer options** screen is hidden by default.

To enable it, navigate to **Settings** | **About phone** and click on the **Build** number seven times. You will find **Developer options** enabled when you return to the previous screen.

Enabling USB debugging

USB debugging enables the IDE to communicate with the device via the USB port. This can be activated after enabling **Developer options** and is done by checking the **USB debugging** using the following this path:

Settings – Developer Options – Debugging – USB debugging

Trusting a computer with installed IDE using secure USB debugging (devices with Android 4.4.2)

You have to accept the RSA key on your phone or tablet before anything can flow between the device via the **Android Debug Bridge** (**ADB**). This is done by connecting the device to the computer via USB, which triggers a notification entitled **Enable USB Debugging?**

Check off **Always allow from this Computer** and click on **OK**.

Setting up the Environment Variables on Windows 7 and higher

Using Ionic and Cordova to build an Android app might require a modification to the **PATH** environment on Windows. This can be done with the following steps:

1. Right-click on **My Computer** and then click on **Properties**.
2. Click on **Advanced System Settings** in the column to the left.
3. In the resulting dialog box, select **Environment Variables**.
4. Select the **PATH** variable and click on **Edit**.
5. Append the following to the **PATH** based on where you installed the SDK:

   ```
   ;C:\Development\adt-bundle\sdk\platform-
   tools;C:\Development\adt-bundle\sdk\tools
   ```

Setting up the Environment Variables for iOS on Mac OS X

Developing for iOS requires you to develop from a machine that runs on Mac OS X. At this point in time, it's not possible to develop iOS applications from Windows. In the following steps, we will outline how to get started with developing Ionic apps for iOS.

Installing the iOS SDK

The following are two ways that can be used to download Xcode:

- From the App Store, search for **Xcode** in the App Store application
- It is available at Apple Developer Downloads, which requires you to register as an Apple Developer.

Command-line tools are integrated within Xcode. Previously, this was installed separately. Once you've downloaded and installed Xcode, you are prepared to handle iOS projects from a machine that has Mac OS X enabled.

MongoDB

In order to store data related to your app and users, your server will need a database—a piece of software that is dedicated solely to data storage and retrieval.

Databases come in many variants. In this book, our focus is **NoSQL** databases, which are so named because they don't use the traditional table-oriented SQL data access architecture that is used by the more well-known relational databases, such as Oracle, MySQL, and PostgreSQL. NoSQL databases are very novel in their design and features and excellent for the tackling of the challenges that one may face in modern app development.

The NoSQL database that we will use throughout this book is MongoDB (it is often abbreviated as MDB or simply Mongo). MongoDB is a document-oriented database that which stores data in documents, which are data structures that are almost identical to the standard JSON format.

Let's have a look at how to install and get MongoDB running. If you have used a more traditional DB system, you may be surprised at how easy it is.

Installation of MongoDB on different Operating System

MongoDB comes in the form of a package for most major OS and versions.

Windows

MongoDB ships with a complete MSI for installation on Windows systems. To download it, go to the project's official website, www.mongodb.org, navigate to **Downloads**, and select the **Windows** tab. You will be offered the following three choices to download:

- **Windows 64-bit R2+**: Use this if you are running Windows Server 2008, Windows 7 64-bit, or a newer version of Windows

- **Windows 32-bit**: Use this if you have a 32-bit Windows installation that is newer than Windows Vista

- **Windows 64-bit legacy**: Use this if you are using Windows Vista 64-bit, Windows Server 2003, or Windows Server 2008.

> MongoDB does not run on Windows XP at all.

After you have downloaded the MSI, run it with administrator privileges in order to perform the installation. The installation wizard will give you a default location where MongoDB will be installed — C:/mongodb/. You can change this if you desire, but it is recommended that you keep it as we will assume that this is the location where MongoDB resides for the remainder of the book.

After the installation has finished, the next step is to configure a data directory where MongoDB can store the data that we will feed it with. The default location for this directory is /data/db. We will need to make sure that this directory exists and is writeable before we start our MongoDB instance for the first time. So, fire up the command prompt with administrator privileges and issue the md/data/dbcommand. After doing so, we are good to start the database server itself. To do so, stay in the command prompt and issue the following command:

```
C:/mongodb/bin/mongodb.exe
```

You should receive a confirmation that MongoDB is now running and listening for connections. All is set!

Linux

On Linux, you will find MongoDB ready-packaged on most major distributions. However, we strongly recommend that you use the project's own repositories in order to make sure that you always have access to the most current security and stability updates.

Ubuntu

First off, you will need to enable the official MongoDB repository. To do so, open a terminal and first import the project's public **GNU Privacy Guard (GPG)** key as follows:

```
sudo apt-key adv --keyserverhkp://keyserver.ubuntu.com:80
--recv7F0CEB10
```

Once this is done, create a listing for the repository itself by issuing the following command:

```
echo 'deb http://downloads-distro.mongodb.org/repo/ubuntu-upstart
dist10gen' | sudo tee /etc/apt/sources.list.d/mongodb.list
```

Your repository listing is now active. Let's make **Advanced Package Tool (APT)** aware of it in order to install MongoDB, as follows:

```
sudo apt-get update
```

Finally, issue the following command to install MongoDB:

```
sudo apt-get install mongodb-org
```

Fedora/RHEL/CentOS

Our first order of business here is to enable the official MongoDB repository. To do so, first make sure that you have the nano text editor installed by opening a terminal and issuing the following command:

```
sudo yum install nano
```

After this is done, add the repository by issuing the following command:

```
sudonano /etc/yum.repos.d/mongodb.repo
```

Nano will open a new, blank text file. Copy and paste the following into the file:

```
[mongodb]
name=MongoDB Repository
baseurl=http://downloads-distro.mongodb.org/repo/redhat/os/x86_64/
gpgcheck=0
enabled=1
```

Save and close the file by pressing *Ctrl+O*, followed by the *Enter* key. This is followed by *Ctrl+X*.

Finally, carry out the installation by issuing the following command:

```
sudo yum install mongodb-org
```

Starting MongoDB

After the installation of MongoDB, you will need to start MongoDB as a service in order to get it running. To do so (on all the distros that were previously mentioned), open a terminal and run the following command:

```
sudo service mongodb start
```

It is important that if you have SELinux running, you must make sure that it allows MongoDB to access its default port. To do so, issue the following before you issue the preceding command:

```
sudosemanage port -a -t mongod_port_t -p tcp 27017
```

Mac OS X

The easiest way to both install and stay up to date with MongoDB on OS X is by using the **Homebrew** package manager. Even if we just use it to install MongoDB here, you will most likely find it useful later for the installation of other software packages that you may need for your own projects after you finish this book.

Installing Homebrew is simple; just open a terminal and issue the following command:

```
ruby -e "$(curl -fsSL
https://raw.githubusercontent.com/Homebrew/install/master/install)"
```

When this finishes, make sure that the Homebrew package database is up to date, as follows:

```
brew update
```

Finally, install MongoDB by simply issuing the following command:

```
brew install mongodb
```

When the install has finished, we will need to define a data directory in order to give a location for MongoDB to store its data. By default, this directory will be at /data/db. So, unless you specify something else, you will need to make sure that this directory exists and is both writeable and readable by the user running your MongoDB instance. For example, if the user running MongoDB on your system is *john*, you will have to issue the following commands:

```
sudomkdir -p /data/db
sudochmod 0755 /data/db
sudochownmongod:mongod /data/db
```

Now that this is done, it is time to start up MongoDB. To do so, make sure that you are logged in as *john*, open a terminal, and simply issue the following command:

```
mongodb
```

That's it! You should receive a notification that MongoDB has started and is listening for a connection. Your instance is ready to go!

Connecting to MongoDB

In order to read and write from the MongoDB instance, we will first need to connect to it. In fact, MongoDB acts as a server in its own right. It exposes its functionality via a network port on which clients can then connect either through the local machine, or even over the Internet.

> Since this functionality is disabled by default due to it being a shoddy security practice, it will require a special configuration of the operating system that MongoDB is running on. We will not discuss this functionality as it falls outside the scope of this book, but we will refer to the MongoDB documentation for several helpful examples of how to achieve it for a variety of OS.

To connect to a MongoDB instance, you will need at least the following information:

- **The IP address of the instance**: If you are accessing an instance on your local machine, this will be *local host* by default.

- **The port on which MongoDB is listening**: Unless you configure a custom value, this will always default to port *27017*.

- **The database that you are trying to connect to**: Don't confuse this with the MongoDB instance itself. Each MongoDB instance can contain any number of databases, with each belonging to different users. The instance simply manages access to them.

Alternatively, you may also need the following:

- A username and its associated password to grant you access to the instance and any databases therein that you are authorized to interact with.

A very easy way to try out this connectivity yourself and verify that MongoDB works as expected is by using the MongoDB shell, a tool that comes installed with MongoDB itself using the methods that we have described previously. How you access the shell varies depending on your OS. I will show each method in the following section and then give an example of how to use the shell itself since this will be the same on all platforms.

Windows

First, make sure that MongoDB is running by following the process that was outlined previously. After this, issue the following command in your command prompt:

```
C:\mongodb\bin\mongo.exe
```

Linux and OS X

First, make sure that MongoDB is running. Then, open a terminal and issue the following command:

```
mongo
```

Now that our shell is running, let's verify that everything works by creating a database and adding some data to it.

To create a database, issue the following command to the shell:

```
use Fruits
```

This will create a database named `Fruits`, to which we can immediately start adding data. (What, you were expecting more overhead? Not in MongoDB!)

We will not add a collection to our database. A collection is simply a basket of data entries, which are grouped based on some logical characteristic. For example, let's suppose that we want a collection of chewy fruits. We then issue the following command:

```
db.createCollection("Chewy")
```

The shell should respond with the { "ok" : 1 } JSON response that tells us that everything went well. Now, let's add some chewy fruits to our collection, as follows:

```
db.Chewy.insert({"name" : "pear"});
db.Chewy.insert({"name" : "apple"});
```

Even if the naming makes it intuitively clear what is going on in the preceding code, don't worry if you do not understand all the details yet. We will get to this in due course.

Finally, let's make MongoDB show us the chewy fruits that we stored. Issue the following command:

```
db.Chewy.find();
```

The shell will respond with something like the following:

```
{ "_id" : ObjectId("54eb3e6043adbad374577df9"), "name" : "apple" }
{ "_id" : ObjectId("54eb4036cdc928dc6a32f686"), "name" : "pear" }
```

The _id numbers will be different on your system, but you will find that the names are the same. All the fruits are where we want them to be.

Congratulations, you now have a fully working MongoDB setup ready for action!

Summary

By now, you may have perhaps noted the red thread running through all the components that we picked for our toolchain; they are all based on JavaScript. This gives us the ability to write all our logic from top to bottom in one single language rather than using different ones for different components (Objective-C or Java for the client, PHP for the server, SQL for the database, or some other unholy combination). As you will see throughout the remainder of the book, this will make it much easier for us to write concise, focused, and comprehensible code.

Once you go through all the preceding steps, I can safely assume that you have the necessary toolkit and you are ready to tackle the upcoming chapters, where we will have an opportunity to create our very first Ionic app and make it work on both Android and iOS devices.

Once you finish setting up your workspace, I recommend that you spend some time reading the Apache Cordova documentation, especially the documentation related to the different platform dependencies.

In the next chapter, we will perform further preparations and configure persistence with MongoDB as part of our preparations for our very own backend for our Ionic app.

2
Configuring Persistence with MongoDB

In this chapter, we will show you how to configure the persistence layer of your app, which is responsible for the provision of the central data storage and retrieval services. For this, we will use MongoDB, the widely popular NoSQL database, and its associated driver and interface for Node.js.

In this chapter, we'll cover the following topics:

- Configuring documents, collections, and databases
- Connecting to MongoDB using a product order database as an example
- Creating relations between documents
- Querying data and displaying results in the command line

Learning outcomes of MongoDB

After reading this chapter, you will have a solid understanding of the basics of how MongoDB stores data. You will also learn how to run queries against a MongoDB instance in order to store, manipulate, and retrieve data on it. You will also understand how to use the Node.js MongoDB driver for the same ends in order to manipulate your data directly from Node.js.

Finally, you will have a healthy dose of refreshment of memory as regards the internal workings of Node.js in order to help you understand how it interconnects with MongoDB.

An introduction to MongoDB

Let's start with a short but informative tour of MongoDB, which will give you the essential knowledge that you need in order to effectively work with it.

First, let's get a good grasp of how data is organized in a MongoDB instance. This will give us the foundation that is required to understand how storage and retrieval operations work later on.

Documents

MongoDB is a NoSQL **Database Management System (DBMS)**. This means that it eschews the traditional table-based data storage model used by SQL-oriented systems such as MySQL, Oracle, and Microsoft SQL Server. Instead, it stores data as documents, which are data structures that are almost identical to standard JSON objects. For example, a MongoDB document can look like this:

```
{
  "_id" : ObjectId("547cb6f109ce675dbffe0da5"),
  "name" : "Fleur-De-Lys Pharmacy",
  "licenseNumber" : "DL 133",
  "address" : "430, Triq Fleur-de-Lys",
  "geolocation" : {
  "lat" : 35.8938857,
  "lng" : 14.46954679999999
  },
  "postCode" : "BKR 9060",
  "localityId" : ObjectId("54c66564e11825536f510963")
}
```

This document represents a pharmacy, with some basic information such as the name, address, and national license number. If you are familiar with JSON, you will feel right at home; this is the standard object notation. However, note an unusual datatype in here—the `ObjectId`. This is a built-in datatype in MongoDB, and it is the default method that is used to uniquely identify a single document. Every single document you store in a MongoDB database is guaranteed to have a unique `_id` member with respect to that database.

> If you are familiar with SQL, you may be tempted to think about it as a column ID. Don't! An `_id` uniquely identifies a document in the entire database, whereas an SQL column ID only uniquely identifies a row in a table.

Collections

Even though you can uniquely identify a document by its _id, life would be a lot simpler if we could somehow organize documents according to some common characteristics. This is where the concept of a collection comes into play. Simply put, a collection is nothing more than a group of documents that exist in a common folder. For example, we can have a collection named Pharmacies, which will store documents like our preceding example.

If you are used to SQL, you may instinctively feel that the documents in the same collection must somehow have the same structure, just like rows in an SQL table do. Surprisingly, this is not even remotely true. Collections only group documents; they do not impose any structural demands on them (apart from the need to have an _id, but this holds for all the documents and has nothing to do with a particular collection). This means that in the collection that we store our pharmacy-related data in, we may also store documents that describe fruit, people, cars, or movies. Whether we should do so is left entirely up to the programmer. This great freedom of structure is one of the most powerful aspects of MongoDB and a key factor that sets it apart from the more traditional DBMS.

Databases

We now know that MongoDB stores data as documents in collections. The last storage concept that we need to mention is the database itself. Simply put, a database in MongoDB is a top-level organizational structure, which holds a group of collections along with information about users who may access the database, security settings, optimizations, and other configuration options. A single MongoDB instance can manage as many databases as server resources will allow.

> It is easy to be misled into thinking that MongoDB itself is the database. Rather, MongoDB is a DBMS, which can manage an arbitrary number of databases.

An example – a product order database

Let's put what we have learned so far into practice and construct a simple MongoDB database that contains data about products, customers, and the orders that the customers have made for specific products. If you are accustomed to other DBMS such as MySQL, you may be surprised to see how simple and intuitive the process is.

Connecting to MongoDB

In order to be able to interact with a MongoDB instance, we first need to be sure that our server is running it. Then, we can access it through the Mongo shell application. In *Chapter 1*, *Setting Up Your Workspace*, we covered in some detail how to install and get MongoDB running on your specific operating system. You should go through these steps if you have not done so already. Once you have verified that MongoDB is running, open the MongoDB shell for your operating system.

Linux and Mac OS X

Start a console and run the following:

```
mongo
```

Windows

Start your command prompt and run the following:

```
C:\mongodb\bin\mongo.exe
```

You will see a prompt starting with the > character. From here, we can issue commands to MongoDB interactively and read the resulting output.

Creating a database

Let's start by defining the database that we are going to work with. In your shell, execute the following:

```
> use OrderBase
```

This will ask MongoDB to switch to a new database, called OrderBase, that we wish to run the commands against. The response will be as follows:

```
switched to db OrderBase
```

But wait, how can we switch to a database that does not exist yet? MongoDB flexibility to the rescue! When you tell MongoDB to use a database, it will create that database for you automatically before switching to it.

Creating our collections

Now that we have created a database, let's populate it with some collections by performing the following steps:

1. Run the following to create a collection for `Products`:

   ```
   > db.createCollection('Products')
   ```

   MongoDB will respond with the following:

   ```
   { "ok" : 1 }
   ```

   The preceding code indicates that the command was executed successfully. Note that the response is returned to us in the JSON format.

2. Let's pause for a minute and break down the preceding command so that we understand what we just did:

   - The `db` is a JavaScript object that represents the currently selected database. In our case, it is `OrderBase`.
   - The `createCollection('Products')` function is one of the many member methods of `db`. Needless to say, it creates a new collection and adds it to `db`. Its parameter, a string, is the name of the new collection.

In other words, working with MongoDB is actually a matter of issuing commands in pure JavaScript. Not only that, but the data itself and the responses to the commands are encoded as JSON! It's obvious right away why MongoDB makes a perfect, seamless fit for JavaScript projects.

1. Let's create two other collections as well to store our orders while we are at it:

   ```
   > db.createCollection('Orders')
   > db.createCollection(Customers'Customers')
   ```

   You will get the same **ok** responses as before.

2. Now, let's add some products to our `Product` collection. In our case, let's say that a product has the following defining characteristics:

   - A name of the string type
   - A price of the float type

We can represent this as a simple JSON object, as follows:

```
{
  "name" : "Apple",
  "price" : 2.5
}
```

3. Inserting name and price into the `Products` collection is equally simple:

```
> db.Products.insert({"name" : "Apple", "price" : 2.5})
```

The response will be as follows:

```
WriteResult({ "nInserted" : 1 })
```

The preceding result contains a `WriteResult` object, giving details about the outcome of a write operation against the MongoDB instance. This particular `WriteResult` instance tells us that the write was successful (as no error was returned), and that we inserted a total of one new document.

4. Again, let's take a closer look at the command that we just issued:

   ◦ The `db` is still the database that we are operating on, which is `OrderBase`.

   ◦ `Products` is our products collection that belongs to `db`.

   ◦ The `insert()` method belongs to the products collection (note that even collections are represented as plain JavaScript objects with properties and methods). It takes a JSON object, such as the one that we defined in the preceding code, and inserts it into the collection as a new document.

   Now that one of our collections actually contains a document, we can ask MongoDB to tell us what is in it.

5. Issue the following command:

```
> db.Products.find()
```

The `find()` method tells MongoDB to look up in the documents from the associated collection. If you pass no parameters to it (an empty find), it will return all the documents in the collection. Fortunately for us, we do not have enough documents (yet) to cause too much screen-scrolling from doing so:

```
{ "_id" : ObjectId("54f8f04a598e782be72d6294"),
  "name" : "Apple",
"price" : 2.5 }
```

This is the same apple that we inserted earlier...or is it? Note that MongoDB created an `ObjectId` instance for it and automatically added it to the objects members. This will always be done (unless you specify a manual _id), since all the documents in a MongoDB database are required to have their own unique _id.

> If you are running this example on your own machine, you will quickly note that the _id values for your objects will differ from the ones seen here since the IDs are randomly generated at the time of insertion.

6. Let's go ahead and insert two more products. However, rather than executing one `insert` statement for each of them, we can instead perform a bulk insertion this time by passing all the objects that we want to insert in a JSON array, as follows:

```
> db.Products.insert([{"name" : "Pear", "price" : 3.0},
{"name" : "Orange", "price" : 3.0}])
```

The response will be as follows:
```
BulkWriteResult({
    "writeErrors" : [ ],
    "writeConcernErrors" : [ ],
    "nInserted" : 2,
    "nUpserted" : 0,
    "nMatched" : 0,
    "nModified" : 0,
    "nRemoved" : 0,
    "upserted" : [ ]
})
```

This response, a `BulkWriteResult` method, is clearly a lot more complex than an ordinary `WriteResult`. We do not need to concern ourselves with what its properties mean just yet. It is enough that we can read from it that two documents were written to the database (`"nInserted" : 2`).

7. Let's issue another `find()` method to make sure that our database contains what we expect:
```
{ "_id" : ObjectId("54f8f04a598e782be72d6294"), "name" :
"Apple", "price" : 2.5 }

{ "_id" : ObjectId("54f8f6b8598e782be72d6295"), "name" :
"Pear", "price" : 3 }

{ "_id" : ObjectId("54f8f6b8598e782be72d6296"), "name" :
"Orange", "price" : 3 }
```

8. Now, let's wrap up by adding some customers as well. We will add our orders a bit later:

```
> db.Customers.insert(
[
{"firstName" : "Jane", "lastName" : "Doley"},
{"firstName" : "John", "lastName" : "Doley"}
])
```

9. Finally, verify that we now have customers to work with by executing the following command:

```
> db.Customers.find()
```

The response will be as follows:

```
{
"_id" : ObjectId("54f94003ea8d3ea069f2f652"),
"firstName" : "Jane",
"lastName" : "Doley"
},
{
"_id" : ObjectId("54f94003ea8d3ea069f2f653"),
"firstName" : "John",
"lastName" : "Doley"
}
```

Creating relations between documents

We now know how to create documents in the collections of a database. However, in real life, it is usually never enough to simply have standalone documents. We will also want to establish some kind of relations between the documents.

For example, in our database, we store information about customers and products, but we also want to store information about orders, which essentially are bills of sale stating that customer *X* has ordered product *Y*.

Let's say that *Jane* wants to order an *Pear*. To achieve this, we could let our orders look like this:

```
{
"customer" :
{
"firstName" : "Jane",
"lastName" : "Doley"
```

```
},
"product" :
{
"name" : "Pear",
"price" : 3
}
}
```

However, the disadvantages of this become clear immediately. It leads to massive data bloating, since the same customer or product can occur in several orders. Hence, its data will need to be repeated in each of the orders. It also makes maintenance a nightmare. If we want to update, say, the price of a product, we need to comb through the database for every single instance where that product appears and make the change.

A much better approach, as recommended by the MongoDB developers, is to use manual references. In this approach, we only store the _id of the document that we wish to refer to rather than the full document.

> There are alternative methods built into MongoDB, but generally, they deal with corner cases and are not optimal for general use. Throughout this book, we will only use the method described here.

We then let the application accessing the database retrieve information about the other document(s), which are referred to as needed. Going back to our order example, this means that the final order document will instead look like this:

```
{
"customerId" : ObjectId("54f94003ea8d3ea069f2f652")
"productId" : ObjectId("54f8f6b8598e782be72d6295")
}
```

Note that we appended Id to the property names in the preceding code. This is a normal convention when dealing with references to other documents, and it is highly recommended that you follow it.

As we have come to expect from MongoDB by now, inserting this new document is no harder than the following:

```
db.Orders.insert({
"customerId" : ObjectId("54f94003ea8d3ea069f2f652"),
"productId" : ObjectId("54f8f6b8598e782be72d6295")
})
```

We can then run `db.Orders.find()` to assure ourselves that everything went as expected:

```
{
"_id" : ObjectId("54f976ccea8d3ea069f2f654"),
"customerId" : ObjectId("54f94003ea8d3ea069f2f652"),
"productId" : ObjectId("54f8f6b8598e782be72d6295")
}
```

It is important to note that even though our order serves no other purpose but to tie two other documents together, it still has its own unique ID.

That's it! We have now constructed a simple database for the storage of information about customers, products, and orders. Next, we will learn how to query it in order to retrieve data for it.

Querying MongoDB

We are now familiar with the overall structure of data storage in MongoDB as well as how to insert and perform some rudimentary retrieval using the `find()` method. Here, we will look at the more advanced usage of `find()` in order to do some more fine-grained data retrieval.

Searching by ID

One of the most common operations on a MongoDB instance is lookups based on ID. As you may recall, every document in a database has a unique _id field, and MongoDB makes it easy to find documents using it.

Let's try this out! Start your Mongo shell and open the OrderBase database again. If you closed it after the last example, you can reopen the database by issuing the following command:

```
> use OrderBase
```

Once the database has been selected, let's say that we want to look up a given product by ID. We select an ID from the earlier example at random and see what we end up with. Remember that the ID will be different on your own machine. So, you will need to select the one that is associated with your own objects:

```
> db.Products.find(
{
_id: ObjectId("54f8f6b8598e782be72d6295")
})
```

The response that we will get for our example is as follows:

```
{ "_id" : ObjectId("54f8f6b8598e782be72d6295"), "name" : "Pear",
"price" : 1.5 }
```

Sure looks like our pear! Now, let's backtrack a bit and see how the lookup works.
Note that we essentially did the same thing as we did when we wanted to see all
the available `Products`:

```
db.Products.find()
```

However, we qualified what we want to find this time by passing a parameter to
`find()`. As we have grown accustomed to this process by now, the parameter, like
most things in MongoDB, is just in JSON:

```
{ _id: ObjectId("54f8f6b8598e782be72d6295") }
```

What we do through this parameter is tell MongoDB that we want to find all
the documents in the `Products` collection whose `_id` property is equal to the
corresponding value in our JSON parameter, which is `ObjectId("54f8f6b8598e782b
e72d6295")` in this case.

Note that the `find()` method can return several results. When searching for an ID,
this makes little sense, since only one ID can belong to any given document and
as such, there can be at the most one result. To accommodate situations like this,
MongoDB provides another method for collections — `findOne()`. It works like find(),
with the sole exception being that it returns at most one result, as follows:

```
> db.Products.findOne({
_id: ObjectId("54f8f6b8598e782be72d6295")
})
```

Searching by property value

We have seen how easy it is to find documents by ID, and it should come as no
surprise that searching by general property values is equally simple. For example,
let's say that we want to find all the products with the name `Orange`. We can do the
following:

```
db.Products.find({"name" : "Orange"})
```

MongoDB will return the following result:

```
{ "_id" : ObjectId("54f8f6b8598e782be72d6296"), "name" : "Orange",
"price" : 3 }
```

In some cases, several documents in a collection will have the same value for the property that we are searching for. In that case, MongoDB will return all the matching ones. Here's an example:

```
db.Products.find({"price" : 3.0})
```

This will return all the products with a price of 3.0. In our case, it will return the following result:

```
{ "_id" : ObjectId("54f8f6b8598e782be72d6296"), "name" : "Orange",
"price" : 3 },
{ "_id" : ObjectId("54f9b82caf8e5041d9e0af09"), "name" : "Pear",
"price" : 3 }
```

Advanced queries

What we have covered here barely scratches the surface of everything that you can possibly do with find(), but it is all that we need to know to be able to configure a basic persistence layer. Throughout the remainder of this book, we will incrementally introduce more advanced queries as the need arises.

Connecting MongoDB and Node.js

We now have a solid understanding of the most basic concepts of how MongoDB works, and it is high time we put them to good work by integrating MongoDB with Node.js. In this section, we will cover this process step-by-step, and see how we can interact with the MongoDB databases directly from within a running Node.js instance.

Setting up a basic project

Before we start, please make sure that you have Node.js and **Node Package Manager** (**NPM**) installed on your system. Refer to *Chapter 1, Setting Up Your Workspace*, for the steps for your particular operating system.

Once you are set, start off by creating a basic project to experiment a bit with the MongoDB instance. Create a folder somewhere and call it MongoNode. Next, open a terminal (OS X/Linux) or the command prompt (Windows), navigate to this folder, and issue the following command:

```
npm init
```

This will launch an interactive wizard for the bootstrapping of a basic Node.js application. In the following code, we provide some standard answers to the questions that the wizard will ask:

```
name: (MongoNode)
version: (0.0.0)
description: Simple project demonstrating how to interface with a
MongoDB instance from Node.js
entry point: (index.js)
test command:
git repository:
keywords:
author: Yours Truly
license: (BSD)
About to write to /home/user/IdeaProjects/nodebook-
ch2/MongoNode/package.json:

{
  "name": "MongoNode",
  "version": "0.0.0",
  "description": "Simple project demonstrating how to interface
  with a MongoDB instance   from Node.js",
  "main": "index.js",
  "scripts": {
  "test": "echo \"Error: no test specified\" && exit 1"
  },
  "author": "Yours Truly",
  "license": "BSD"
}

Is this ok? (yes)
```

Once the bootstrapping finishes, create a new file named index.js. Open it in your favorite text editor and type the following:

```
console.log('Hello World!');
```

Save the file and then open a terminal. Navigate into the folder that we just created and run the following command:

```
node index.js
```

You should get the following familiar output:

Hello World!

We are now assured that Node.js works as expected. So, let's go ahead and see how we can get in touch with the database that we constructed earlier.

Connecting to MongoDB

Now, let's set up the bare metal to interface with a MongoDB instance. The first thing that we will need to do is install the official MongoDB driver for Node.js. Inside your project folder, issue the following command in the terminal:

```
npm install mongodb -save
```

This will make npm carry out the complete installation procedure. Once this is done, we will have all the basic functionalities that we need to interact with the MongoDB instance.

After the install finishes, create a new file named database.js, open it in your text editor, and insert the following. Don't worry if it is quite a lot of code as compared to what we have seen so far; I added quite a lot of commentary to explain what is going on:

```javascript
// Our primary interface for the MongoDB instance
var MongoClient = require('mongodb').MongoClient;

// Used in order to verify correct return values
var assert = require('assert');

/**
 *
 * @param databaseName - name of the database we are connecting to
 * @param callBack - callback to execute when connection finishes
 */
var connect = function (databaseName, callback) {

  // URL to the MongoDB instance we are connecting to
  var url = 'mongodb://localhost:27017/' + databaseName;

  // Connect to our MongoDB instance, retrieve the selected
  // database, and execute a callback on it.
  MongoClient.connect(url, function (error, database) {

    // Make sure that no error was thrown
```

```
    assert.equal(null, error);

    console.log("Successfully connected to MongoDB instance!");

    callback(database);
  });
};

/**
 * Executes the find() method of the target collection in the
 * target database, optionally with a query.
 * @param databaseName - name of the database
 * @param collectionName - name of the collection
 * @param query - optional query parameters for find()
 */
exports.find = function (databaseName, collectionName, query) {
  connect(databaseName, function (database) {

    // The collection we want to find documents from
    var collection = database.collection(collectionName);

    // Search the given collection in the given database for
    // all documents which match the criteria, convert them to
    // an array, and finally execute a callback on them.
    collection.find(query).toArray(
      // Callback method
      function (err, documents) {

        // Make sure nothing went wrong
        assert.equal(err, null);

        // Print all the documents that we found, if any
        console.log("MongoDB returned the following
        documents:");
        console.dir(documents);

        // Close the database connection to free resources
        database.close();
    })
  })
};
```

Next, let's import the database module in the `index.js` file. Remove everything from this file and insert the following in it:

```
var database = require('./database');
```

This will allow us to use our database interface like a regular Node.js module.

Finally, let's give it a run and make sure that everything works. Insert the following line in the `index.js` file:

```
database.find('OrderBase', 'Products', {});
```

The preceding command should immediately seem familiar to you; it is the same as when we ran the following command in our earlier example:

```
db.Products.find();
```

Here, we simply wrapped the parameters in logic so that it can be run in the Node.js instance.

To run the Node.js instance, issue the following command in your terminal again:

```
node index.js
```

You should receive something like the following:

Successfully connected to MongoDB instance!

MongoDB returned the following documents:

```
[ { _id: 54f8f04a598e782be72d6294, name: 'Apple', price: 2.5 },
  { _id: 54f8f6b8598e782be72d6295, name: 'Pear', price: 1.5 },
  { _id: 54f8f6b8598e782be72d6296, name: 'Orange', price: 3 },
  { _id: 54f9b82caf8e5041d9e0af09, name: 'Banana', price: 3 } ]
```

Summary

That's it! You just wrote your first Node.js app by making use of MongoDB! It is very simple. Note that as we go ahead, we will structure our code a bit differently, but now, you have the basic know-how that is needed to make it work. Next, we will study some advanced topics and look at how we can use Node.js and MongoDB in order to construct a full-fledged API.

With this chapter, we provided you with the basic knowledge that is needed to start building your very own database for your Ionic mobile application, which is one of the first stepping stones in building your very own backend for your cross-platform mobile application.

As we move further along, you will realize that the essentials that we learned through this chapter will provide us with the necessary knowledge that is required to start building our own API, which will be done in the next chapter.

3
Creating an API

No matter how sophisticated your backend is, it will most likely be of no use unless you can make its services available to your clients in some way. The most common way to do so is through an **Application Programming Interface (API)** — a set of well-defined access methods for your backend. Through this, you allow clients to request data, perform calculations, and request other services offered by it.

In this chapter, we will look at how to construct such an API according to the widely used **Representational State Transfer (REST)** architecture. We will cover the theoretical basics of REST and then see how to implement its condensed version using Node.js according to our needs. We will round up by accessing the API via a REST client in order to see how it works in action.

Learning outcomes of the RESTful API

After reading this chapter, you will understand what a RESTful API is and how it is structured. You will also have gained a thorough understanding of how to define data access points in Node.js and work with these through a REST client.

RESTing easy

One of the greatest challenges of the Internet age has always been to make networked services talk to each other in a uniform and efficient manner. It is the reason behind why we have developed a plethora of communications protocols that we depend on today, such as the all-important TCP/IP stack. Protocols like these make formal communication between applications a straightforward process, at least as far as the actual bytes on the wire are concerned.

However, there is no single de facto specification for how applications should communicate data abstractions to each other. Raw TCP/IP only understands the exchange of data packets; it knows nothing about abstractions such as customers, orders or products. To raise the abstraction level and build an interface that allows intuitive communication with our backend, we will have to rely on a custom implementation of one or more architectural patterns in order to get what we want. Today, there are several such patterns in wide usage. You may already be familiar with terms such as SOAP, WSDL, and perhaps even our target here—REST.

It's all hypermedia

REST is an architectural pattern where two or more applications exchange resources among themselves through a set of operations on these resources. The resources are sets of data types that all the involved applications, such as products, customers, and so on, know about. For example, a client application can either ask a server application to give it a list of all the resources of a given kind that it stores, or ask the server to register a new instance of a given resource in its database. All such operations are communicated only by using the standard HTTP protocol, which makes the process both intuitive and easy to implement.

At the heart of the RESTful communication are the common HTTP verbs—**GET**, **POST**, **PUT**, and **DELETE**. In terms of the popular **CRUD (Create/Read/Update/ Delete)** acronym, C corresponds to **POST**, **R** to **GET**, **U** to **PUT**, and finally **D** to, well, **DELETE**. If you are familiar with HTTP, you will already know that these verbs represent different request types sent between two HTTP-speaking applications. For example, when you type `www.google.com` in the URL field of your browser and press the *Enter* key, the browser will issue a GET request to the server the URL is bound to, asking it to return whatever data is located at server. Typically, this data is in a standard format such as **HTML**, **XML**, or **JSON**. In the same way, when you fill out a form on a webpage and press the **submit** button (or its equivalent), a request is sent to the server through a POST request, which carries the request data that you entered for the server to process.

As mentioned before, REST functions by using these verbs to communicate operations on various resources that the involved parties know about. For example, in our case, we may want to tell a server the following by using the RESTful requests:

- Fetch all the products that are available on the server (the verb is GET)

- Fetch the product with an ID of abcd1234 (the verb is GET)

- Place a new order with a product named abcd1234 for a customer with an ID of xyz456 (the verb is POST)

- Update the price of the product with an ID of `abcd1234` to 500 million dollars (the verb is `PUT`)

- Delete the product with an ID of `abcd1234` since nobody is buying it anymore (the verb is `DELETE`)

By convention, REST uses the following common base URL structure for requests operating on a given resource (elements in the brackets are optional):

```
http://<domain>/[api name]/[api version]/<resource>
```

For our product example, a RESTful base URL for this resource is as follows:

```
http://myserver.com/myapi/v1/products
```

In the following section, we will demonstrate how such actions are carried out in practice using the HTTP verbs.

GET

In the context of REST, a GET request always indicates a retrieval operation. Thus, we say that the GET requests are the only non-mutating ones among the common verbs, since they do not change the state of the associated resource on the server.

In REST, there are two standard GET operations that any API should ideally implement:

GET all

The following are the key features of the GET all operation:

- It sends a blank GET request to the base URL for the resource

- It returns all the resources of a given type

- An example of resources that are given by the GET all operation is `http://myserver.com/myapi/v1/products`.

GET by ID

The following are the key features of the GET by ID operation:

- It returns the resource with the specific ID

- An example of this type of operation is `http://myserver.com/myapi/v1/products/abcd1234` (using the path parameters) or `http://myserver.com/myapi/v1/products?id=abcd1234` (using the query parameters)

Whether you should use path parameters or query parameters is entirely up to you, and it is not mandated by the REST conventions. Here, and for the remainder of the book, we will use query parameters, since this is the normal HTTP way of doing things and a bit easier to understand and implement.

POST

POST requests are used in order to create new instances of a given resource. Normally, a conventional REST server will provide documentation about the fields of the resource that you need to specify for the creation to succeed.

PUT

PUT is used in order to create or update a resource. It works almost identically to POST, with the exception that if you supply a resource ID with your request, the server will first find that specific resource and then replace each field of that resource with the equivalent field in your request.

For example, consider a situation where your server has a resource of the product type, as follows:

```
{
   name: 'Apple',
   price: 50,
   id: 'abcd1234'
}
```

Let's suppose that you submit a PUT request with the following form data:

```
{
   id: 'abcd1234'
   price: 500000000,
}
```

The same resource will have the following state on the server after the transaction concludes:

```
{
   name: "Apple",
   price: 500000000,
   id: "abcd1234"
}
```

DELETE

DELETE is used in order to, well, delete a resource on the server. All you need to do is supply the resource ID in your request. For example, sending a DELETE request to `http://myserver.com/myapi/v1/products/abcd1234` will delete the product with an ID of `abcd1234` on the server.

Building a RESTful API with Node.js

Now that we have covered the basics of REST, let's put it into practice and build an API for `OrderBase`, which was constructed in the previous chapter. If you have not done so already, please take a moment to review the code that we wrote there in order to make sure that you understand what happens between our API and the database in this example.

Setting up the RESTful API

Start with creating a workspace for our server. On your drive, create a folder named `order_api`, step into this folder, and create and execute the file named `api.js`. Finally, open a terminal and execute the following:

```
npm init
```

As we saw in the previous chapter, this will give you a few questions to answer in order to bootstrap the Node.js server. When the questions ask you for the entry point, be sure to specify `api.js`, since this is the main file that your server configuration will be read from.

Next, you will need to import the database interface module that we created in *Chapter 2, Configuring Persistence with MongoDB*. To do so, first install the `mongodb` driver:

```
npm install mongodb
```

Then, you can import the module itself in two ways:

- Copy and paste the `database.js` file from the previous chapter into the current directory and add `var database = require('./database');` to the top of your `api.js` file

- Add `var database = require([pathToDatabase])` to the top of your `api.js` file, where `[pathToDatabase]` is the full system path to your `database.js` file

Once this is done, open the `api.js` file. Let's start adding some code for our API.

The HTTP module

The first thing we will need is a way to actually open up the Node.js instance to the network and enable it to communicate over the HTTP protocol, since this will be the core driver of our API's functionality.

In order to achieve this, we will include the standard HTTP module in our server. Add the following line to the top of your `api.js` file:

```
var http = require('http');
```

This will cause Node.js to load the HTTP module, a powerful component that can be used to listen for and process HTTP requests as well as send responses to the clients.

Now, with the module in place, let's hotwire Node.js to start listening and respond to simple HTTP requests. Add the following to your file:

```
var server = http.createServer(function (req, res) {
    res.writeHead(200);
    res.end("I am a fledgling API, and I am alright");
});
server.listen(8080);

console.log('Up, running and ready for action!');
```

That's it! If this is your first time using the HTTP module, you may be surprised at how simple this setup is. It is not everyday that you write a fully functioning HTTP server in seven lines of code! Node.js is just that good.

Let's give the server a run to make sure that it is working alright. Open your favorite browser and navigate to `http://localhost:8080`. You will see the following line of text:

I am a fledgling API, and I am alright

All is well. We are now ready to start making our API do something more interesting than just show the same text over and over. However, let's take a closer look first at how the HTTP module actually works and services requests.

Dissecting the HTTP server

Looking at our server code, we are really just doing the following two things:

1. Configuring the event loop for our server is what it should do whenever an HTTP request comes in. This is done by invoking the `http.createServer()` method, which takes a callback function as a parameter, which will be executed for each incoming request.

2. Bind the server to a given network port in the host machine and start listening for incoming connections on that port.

The interesting bit of the first item is the callback function:

```
function (req, res) {
  res.writeHead(200);
  res.end("I am a fledgling API, and I am alright");
}
```

This method takes two arguments, `req` and `res`. As you may have guessed, they refer to the HTTP *request* and the associated *response*. The `req` parameter will contain all the data associated with the incoming HTTP request, such as origin, headers, payload, cookies, and more. The `res` parameter is the HTTP response, which will be emitted back to the caller when the method is returned.

You may wonder why the response is passed as a parameter to a function that obviously handles incoming requests. This is a matter of design. The `res` parameter is actually created outside the function and passed to it so that you can do what modifications you see fit to it before the HTTP module takes control again, finalizes it, and sends it back to the sender.

In our function, we do only the following two things:

- We set the response code of the response to *200*, indicating a successful request cycle.
- We append a string to the body (that is, the payload) of the response—"`I am a fledgling API, and I am alright`".

That's it as regards handling and responding to requests (really!).

Let's put this to use and start returning something more interesting.

Returning JSON

Normally, REST APIs will support serving data in several different formats, such as JSON and XML. For the sake of simplicity, we will only focus on JSON here. This makes sense in the context of what we have seen so far, where everything high and low is JavaScript - and JSON-oriented anyway.

Thankfully, returning a JSON object to our caller is almost trivial; we just need to make a few adjustments inside our callback function:

- Specify the content type of the response as JSON
- Convert the JSON object that we want to send back to a string

The first adjustment is done by modifying the `Content-Type` header of our response. In your code, you have the following code line:

```
res.writeHead(200);
```

You can change this to the following code line:

```
res.writeHead(200, {'Content-Type': 'application/json'});
```

This additional parameter, passed to the `writeHead()` method, is a JSON object with custom values for headers in the response object. If you don't specify headers, the HTTP module will generally set sensible defaults, but you should always be explicit when you are certain about what a header should be set to. Here, we want to make it clear to the client that we are sending them a JSON object as a response, and we set the `Content-Type` header accordingly.

To address the second item, let's first add a JSON object to send back to the client. After the `res.writeHead()` method, add the following:

```
var myProduct = {
  name: 'Apple',
  price: 600
};
```

Next, we need to turn this JSON object into a string in order to package it into the response. To do so, we can use the native Javascript `JSON.stringify()` method. As expected, this method takes a JSON object and returns a string representation of that object. Modify the following line:

```
res.end('I am a fledgling API, and I am alright');
```

Change the preceding line to the following:

```
res.end(JSON.stringify(myProduct));
```

We're done! Save your changes, restart the Node.js instance (just close and start it again), and refresh your browser window for the server. You will see the text:

```
{
  'name':'Apple',
  'price':600
}
```

We now have a full-fledged, JSON-serving HTTP server ticking. It's about time that we got down to the serious stuff.

Implementing our GET handlers

Let's begin by implementing basic GET methods for our resources. You may recall that we mentioned before that a good REST API should at least implement two of them — GET by ID and GET all. Since we like to be standards-compliant, that is what we will use here.

Implementing a router

Our first order of business is to provide a way for our Node.js instance to differentiate between the different URLs that it receives requests for. Until now, our server only had to handle requests to its root URL (`http://localhost:8080/`), but in order to do something more interesting, we want to be able to generate custom responses for more specific URLs, such as `http://localhost:8080/api/products`.

Fortunately, Node.js again provides an out-of-the-box way to achieve this — the URL module.

Add the following just after the `var http = require('http');` line:

```
var URL = require('URL');
```

This will import the URL module. We can now use it to break down the incoming requests and take action depending on how their URLs are structured.

Modify the `http.createServer()` call to look like this:

```
var server = http.createServer(function (req, res) {

  // Break down the incoming URL into its components
  var parsedURL = URL.parse(req.URL, true);

  // Determine a response based on the URL
  switch (parsedURL.pathname) {
    case '/api/products':
    // Find and return the product with the given id
    if (parsedURL.query.id) {
      findProductById(id, req, res);
    }
    // There is no id specified, return all products
    else {
      findAllProducts(req, res);
    }
    break;
    default:
    res.end('You shall not pass!');
  }
});
```

Note that we introduced two new methods, `findAllProducts` and `findProductById`. These are utility methods, which we will define separately. Along with them, we will define some generic helper methods to help make data access less cumbersome for us. Go ahead and add the following before the `createServer()` call:

```
// Generic find methods (GET)

function findAllResources(resourceName, req, res) {
  database.find('OrderBase', resourceName, {}, function (err,
  resources) {
  res.writeHead(200, {'Content-Type': 'application/json'});
  res.end(JSON.stringify(resources));
  });
};

var findResourceById = function (resourceName, id, req, res) {
  database.find('OrderBase', resourceName, {'_id': id}, function
  (err, resource) {
```

```
    res.writeHead(200, {'Content-Type': 'application/json'});
    res.end(JSON.stringify(resource));
    });
};

// Product methods

var findAllProducts = function (req, res) {
  findAllResources('Products', req, res);
};

var findProductById = function (id, req, res) {
  findResourceById('Products', id, req, res);
};
```

The generic methods are straightforward. They simply make use of the MongoDB interface that we created in *Chapter 2, Configuring Persistence with MongoDB*, in order to retrieve either all the documents from a specific collection, or just a single document by its ID. The specific product methods make use of these generic methods in order to find products in this fashion.

For the sake of brevity, we do not implement similar methods for the customer and order here; they are identical to the ones used for the product. Just change the name of the resource and add appropriate paths inside the createServer() method. You can see the complete example in the source code accompanying the book.

Implementing our POST handlers

We will now move on to adding handlers to create new instances of a resource. To do so, we need to distinguish not only between the URLs, but also between the request types. Modify your createServer() invocation so that it looks like the following:

```
var server = http.createServer(function (req, res) {

  // breaks down the incoming URL into its components
  var parsedURL = URL.parse(req.URL, true);

  // determine a response based on the URL
  switch (parsedURL.pathname) {
    case '/api/products':
```

```
    if (req.method === 'GET') {
      // Find and return the product with the given id
      if (parsedURL.query.id) {
        findProductById(id, req, res)
      }
      // There is no id specified, return all products
      else {
        findAllProducts(req, res);
      }
    }
    else if (req.method === 'POST') {

      //Extract the data stored in the POST body
      var body = '';
      req.on('data', function (dataChunk) {
        body += dataChunk;
      });
      req.on('end', function () {
        // Done pulling data from the POST body.
        // Turn it into JSON and proceed to store it in the
        database.
        var postJSON = JSON.parse(body);
        insertProduct(postJSON, req, res);
      });
    }
    break;
    default:
    res.end('You shall not pass!');
  }
});
```

Note that we have introduced another handler method, `insertProduct()`. We define it, along with its corresponding generic method, like we did before:

```
// Generic insert/update methods (POST, PUT)

var insertResource = function (resourceName, resource, req, res) {
  database.insert('OrderBase', resourceName, resource, function
  (err, resource) {
```

```
    res.writeHead(200, {'Content-Type': 'application/json'});
    res.end(JSON.stringify(resource));
    });
};

// Product methods

var insertProduct = function (product, req, res) {
    insertResource('OrderBase', 'Product', product, function (err,
    result) {
    res.writeHead(200, {'Content-Type': 'application/json'});
    res.end(JSON.stringify(result));
    });
};
```

Again, the implementation of this functionality for the other resources is the same, with the exception of the name. We do not replicate them here.

Implementing the DELETE and PUT handlers

Handling DELETE and PUT is analogous to handling a GET and POST request respectively, with the exception of the method being changed. Thus, we recommend that you refer to the accompanying source code to see the full implementation.

Testing the API

Until now, we have used a normal browser to poke at our API and see what it returns. However, this is far from optimal. Most browsers only make it easy to send GET requests, whereas an HTML form or something similar to it is needed in order to send POST requests. Let's not even get started with the DELETE and PUT requests.

To test the REST API, it is a much better idea to use a dedicated REST client, which will give you more options, make it easier to send requests, and thoroughly analyze responses. A very popular (and free) tool is Postman, which is a Chrome extension. It runs on all major operating systems. You can download it for free from `https://www.getpostman.com/`. The install process is very straightforward, and it will not be covered here.

Once you have installed Postman, start it up. Let's post away at our API. First, let's try asking the backend to send us all the products that it currently stores. Enter the products' root URL in Postman's URL field, make sure that GET is selected among the methods in the combobox to the right, and then click on the **Send** button. You should get something that looks like the following screenshot:

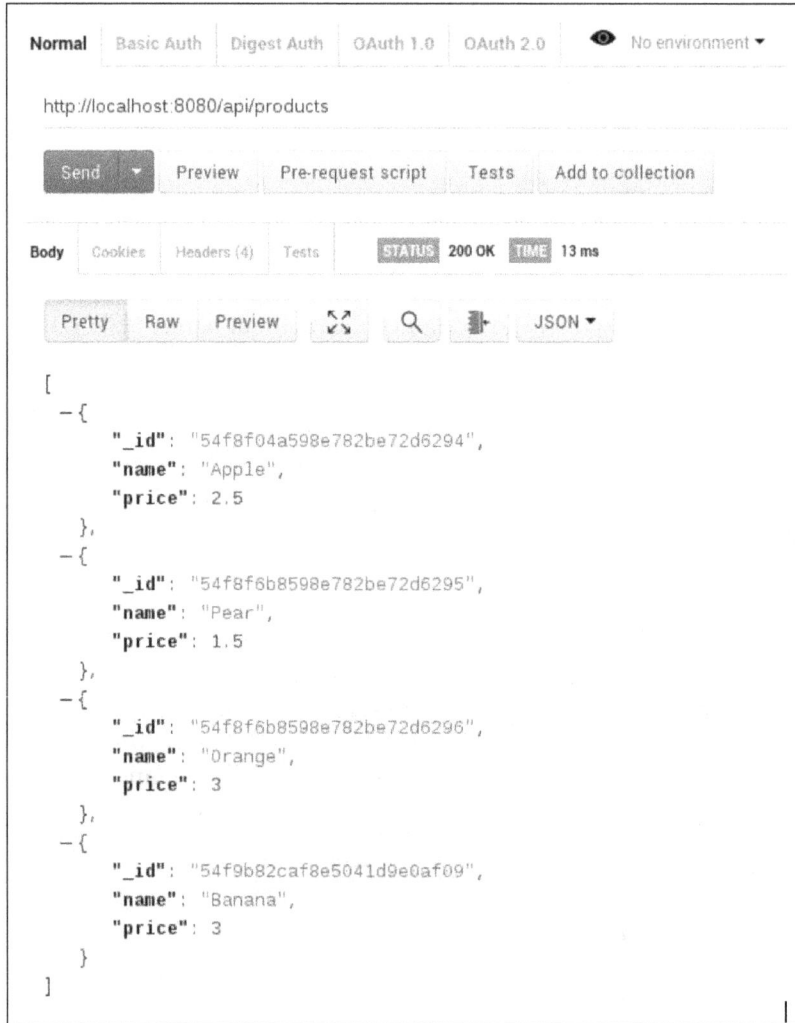

```
Normal    Basic Auth    Digest Auth    OAuth 1.0    OAuth 2.0    ●  No environment ▼

http://localhost:8080/api/products

 Send   ▼     Preview     Pre-request script     Tests     Add to collection

Body    Cookies    Headers (4)    Tests        STATUS  200 OK   TIME  13 ms

 Pretty    Raw    Preview    ⤢    Q    ▐⊩    JSON ▼

[
  - {
        "_id":  "54f8f04a598e782be72d6294",
        "name":  "Apple",
        "price": 2.5
    },
  - {
        "_id":  "54f8f6b8598e782be72d6295",
        "name":  "Pear",
        "price": 1.5
    },
  - {
        "_id":  "54f8f6b8598e782be72d6296",
        "name":  "Orange",
        "price": 3
    },
  - {
        "_id":  "54f9b82caf8e5041d9e0af09",
        "name":  "Banana",
        "price": 3
    }
]
```

Now, let's try to POST a new product to the backend. Keep the same URL, but change the method to POST in the combobox to the right. Next, add some data before sending; select **Raw** from the button group under the URL field and enter the following:

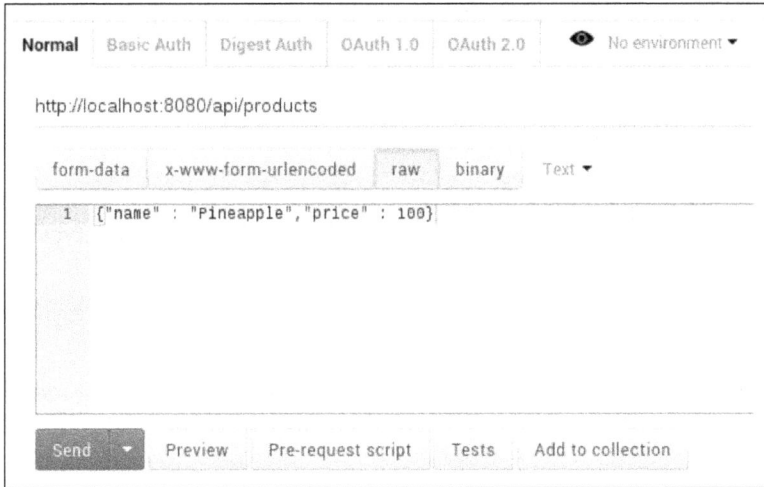

Click on **Send** to fire off the request (note how much easier this is than using a plain browser). Finally, let's pull all the products again in order to make sure that the new product was indeed added:

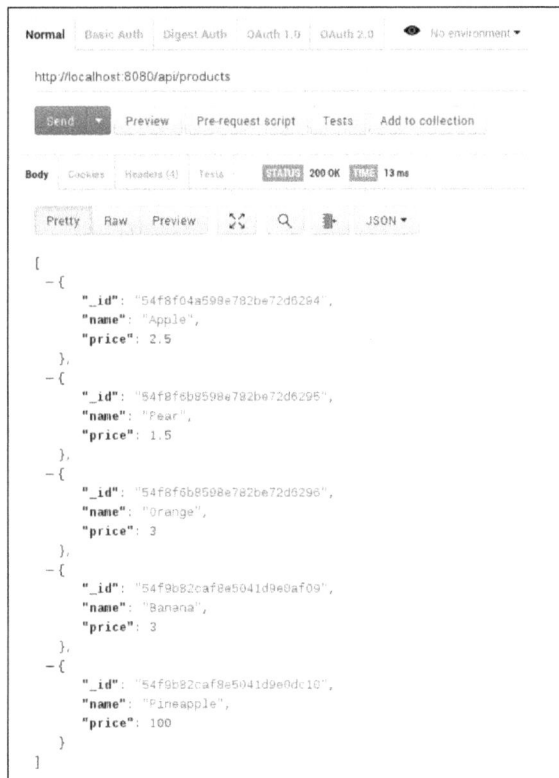

That's it! Our API is working, and we are ready to start moving towards some serious usage.

Moving forward

In this chapter, we studied the bare bones and built our API from scratch by using only the functionality provided by the core Node.js modules themselves. We did this in order to show you how Node.js works in reality and to avoid binding you to any given framework that already does the things that we implemented here, in case you use another one in your own career.

That being said, as an addendum, we would still like to point you to some frameworks that we ourselves recommend in order to build powerful APIs quickly with Node.js:

- **Express.js**: Express is a framework that can be used to build robust, complete web applications using Node.js. It comes with a host of powerful features, including an advanced router, which makes it very easy to handle incoming requests based on the URL (forget about the switch and if statements like we did here), extract, parse, and validate data, connect to external data sources, and much more.

  Express.js is largely seen as the de facto framework for Node.js, and it has a large community and plugins available for it, making it easy to find help and extra functionality as needed.

  > For more information on this framework, visit `http://expressjs.com/`.

- **Loopback.js**: This was developed by the same team that created Express.js. Loopback.js is an Express.js derivative, which is designed solely for the creation of APIs. It comes with a powerful terminal interface, where you can quickly create and modify resources, relations between resources, security, and much more. Loopback automatically generates full RESTful APIs for your resources, which means that you have to write almost no code for cases, such as the examples that we looked at in this chapter.

  > For more information about this framework, visit `http://loopback.io/`.

Summary

You should now have a good understanding of the basic concepts of RESTful APIs as well as how you can implement them using Node.js and access them from a network.

In the next chapter, we will continue improving our API and show how to secure it in order to prevent unauthorized access to your resources.

4
Securing Your Backend

In the previous chapters, we built a rudimentary but functional backend layer by layer to provide basic services for a basic shop-like app. So far, we haven't been paying too much attention to security; everyone with access to the server can execute any command exposed by our API, even if it involves deleting the whole product database!

In this chapter, we are going to remedy this by building a basic security mechanism in order to control user access. Specifically, we will deal with token-based authentication and show you how this makes it easy to limit access to your backend. By doing this, we will introduce the concept of roles and how they figure in our authentication scheme.

Understanding the outcomes of token-based authentication

After reading this chapter, you will understand what token-based authentication is and how it can be used to limit the functionality of an API based on a user's status. You will further understand what roles are and how they affect authentication. Finally, you will know how to implement this authentication mechanism using only the technologies that we have introduced so far.

The theoretical bit

Before we start writing code, let's have an overview of the concepts involved and how they relate both to security and one another.

A small token of trust

Security has always been one of the most pressing concerns in all areas of software development. It is almost never enough to have a system that is fast, scalable, and robust if it doesn't have an adequate mechanism to protect it from malicious users.

In the context of publicly accessible servers such as ours, security is all the more pressing since our API will be exposed to an entire planet of potentially nasty people. Somehow, we need to make sure that the people who request services from it are who they say they are and are allowed to do what they want to do.

A simple yet powerful technique that has emerged to accomplish this is **token-based authentication**. In this, each legitimate user is given an access token (usually a hash), which uniquely identifies the user of a server. The user needs to submit the token along with every request that requires authentication, and the server in turn validates the token in order to determine whether access should be granted.

In order to obtain an access token, the user will first need to initially authenticate themselves to the server in some way. Commonly, this is done via a normal username-password check. If a correctly matching username and password is provided, the server responds by generating an access token, certifying that the user is authenticated to access the server.

Playing your role

In most software systems, not all users are created equal. Some, such as the administrators, are intended to have broader access to the system than the common users. There are several schemes that allow us to limit what functions within a system a user can access, but the most common one probably is to use roles. Put simply, a role is an attribute that grants its holder a certain level of access to the system. For example, a user with the role of **administrator** may have full access to read and write system records, whereas a user with the role of **reader** may just be able to read them. Further more, a user with the role of **BookWorm** may only have access to read data records classified as **books**, and so on.

Putting it all together

Now, it is probably apparent how roles and tokens figure in the authentication scheme that we want to create. The lifetime of an authenticated request will proceed in the following way:

1. The server receives an API request.
2. The server checks whether a token is provided.

If it is not provided, it returns a *403* (that is, **forbidden**).

3. The server checks whether the token is in the database.

   If it is not in the database, it returns a *403*.

4. The server retrieves the user's role.

5. The server verifies that the user's role matches the requirements of the API call.

   If it does not match them, it returns a *403*.

6. The server handles the request and returns an appropriate response to the user.

Implementation

We are now ready to write a functional implementation of the authentication system.

The first thing that we need to do is expand our database to accommodate the necessary documents. In particular, we need to add the following three new collections:

- **Users**: These are the users who can access the server via the API
- **Roles**: These are the roles that can be assigned to users
- **Access Tokens**: These are the access tokens for authenticated users

We will also need to add some rudimentary logic to our API to register users and enable them to log in.

Adding the new collections

Open your MongoDB shell and execute the following:

```
use OrderBase;
db.createCollection('User');
db.createCollection('Role');
db.createCollection('AccessToken');
```

This will create the necessary collections that we need to store users and their roles and tokens. The new documents will have the following structure:

```
User:
{
    firstName,
    lastName,
```

```
        email,
        roleID,
        password
    }
```

For now, we will not add any users or tokens (this comes later when we extend the API), but we will add the roles that we are going to use. To keep it simple, we will just have two of them:

- **Producer**: This is the user who sells goods in the shop and who can add additional products to it.
- **Customer**: This is the user who buys things from the shop and who can create orders and retrieve information about products as well as orders that were created by the current user.

It is understood that default `ObjectID` generated by MongoDB will be included in the preceding code. For the access token entity, we simply use `ObjectID` as the hash of the token, since this value is guaranteed to be unique with respect to the database that we are working with.

Adding an authentication module

To maintain modularity and simplify the authentication process, we will create a separate module to validate the access privileges of a given user.

In your project directory, add the following file named `authentication.js`. Open the file and insert the following:

```
var db = require('./database');

module.exports = {
  database: 'OrderBase',
  collection: 'AccessTokens',
  generateToken: function (user, callback) {
    var token = {
      userID: user._id
    }
  }
}

  // Persist and return the token
  db.insert(this.database, this.collection, token, function (err,
  res) {
    if (err) {
      callback(err, null);
    } else {
```

```
          callback(null, res);
      }
    });
  },
  authenticate: function (user, password, callback) {
    if (user.password ==== password) {
      // Create a new token for the user
      this.generateToken(user, function (err, res) {
        callback(null, res);
    });});
    } else {
      callback({
        error: 'Authentication error',
        message: 'Incorrect username or password'
      }, null);
    }
  }
}
}
```

Next, import the module into your entry module, as follows:

```
var authentication = require('./authentication');
```

Creating functions to register and help users log in

We will need to add endpoints to our API for the purpose of both creating and authenticating users who wish to interact with it. In light of what we have done thus far, this is easy to do.

Registering users

We begin by adding a URL endpoint for adding users. This will be very familiar in terms of what we already did when creating the REST API in the previous chapter; all that we are going to do is create a POST method for the user collection. First, add the following utility method:

```
var insertUser = function (user, req, res) {
  insertResource('OrderBase', 'User', user, function (err, result)
  {
    res.writeHead(200, {"Content-Type": "application/json"});
    res.end(JSON.stringify(result));
  });
};
```

Next, modify your router to include the following `case` statement:

```
case 'api/users/register':
  if (req.method === 'POST') {
    var body = "";
    req.on('data', function (dataChunk) {
      body += dataChunk;
    });
    req.on('end', function () {

      // Done pulling data from the POST body.
      // Turn it into JSON and proceed to store.
      var postJSON = JSON.parse(body);

      // validate that the required fields exist
      if (postJSON.email
      && postJSON.password
      && postJSON.firstName
      && postJSON.lastName) {
        insertUser(postJSON, req, res);
      } else {
        res.end('All mandatory fields must be provided');
      }
    });
  }
  break;
```

This is all we need to register users. Registrations can now be handled through a simple POST request to the /api/users/register endpoint.

Enabling users to log in

To enable users to log in via our API, we will need to accomplish the following three things:

- Make sure that the user exists
- Make sure that a matching password was provided by the the user
- Return an access token, which can be used by the user for future access

Luckily, all but the first of the preceding list are taken care of by the authentication module that we designed earlier. All that we need to do is plug it into our router. To do this, we will also need to design a new endpoint for the login part.

Add the following `case` to your router configuration:

```
case 'api/users/login':
  if (req.method === 'POST') {
    var body = "";
    req.on('data', function (dataChunk) {
      body += dataChunk;
    });
    req.on('end', function () {

      var postJSON = JSON.parse(body);

      // make sure that email and password have been provided
      if (postJSON.email && postJSON.password) {
        findUserByEmail(postJSON.email, function (err, user) {
          if (err) {
            res.writeHead(404, {"Content-Type":
            "application/json"});
            res.end({
              error: "User not found",
              message: "No user found for the specified email"
            });
          } else {
            // Authenticate the user
            authenticator.authenticate(
            user, postJSON.password, function(err, token) {
              if(err) {
                res.end({
                  error: "Authentication failure",
                  message: "User email and password do not match"
                });
              } else {
                res.writeHead(200, {"Content-Type":
                "application/json"});
                res.end(JSON.stringify(token));
              }
            });
          }
        });

      } else {
```

```
      res.end('All mandatory fields must be provided');
    }
  });
}
  break;
```

In the preceding code, we added the following simple method in order to handle the looking up of a user by e-mail:

```
var findUserByEmail = function (email, callback) {
  database.find('OrderBase', 'User', {email: email}, function
  (err, user) {
    if (err) {
      callback(err, null);
    } else {
      callback(null, user);
    }
  });
};
```

That's all we need as far as user management is concerned for now. Now, let's add the finishing touch and set up the actual security for our endpoints.

Extending our API

We are now ready to modify our API in order to add the authentication features that we have developed so far. First, let's determine exactly how the access policies should work:

- **Customers** should be able to create (`insert`) orders and retrieve (`get`) information about products and nothing else
- **Producers** should be able to retrieve information about orders and products and also insert new products

We will accomplish this by placing a simple token and role check on each endpoint. The check will simply verify the following:

- The token is legitimate
- The user associated with the token has the role that is necessary to perform the requested action

To start, we will add a new function to the `authentication` module, which will be responsible for checking whether a given token is associated with a given role:

```
tokenOwnerHasRole: function (token, roleName, callback) {
  var database = this.database;
  db.find(database, 'User', {_id: token.userID}, function (err,
  user) {
    db.find(database, 'Role', {_id: user.roleID}, function (err,
    role) {
      if(err){
        callback(err, false);
      }
      else if (role.name ==== roleName) {
        callback(null, true);
      }
      else {
        callback(null, false);
      }
    });
  });
}
```

This method is all that we need to verify the roles for the token provided (implicitly checking whether the user who owns the token has the specified role).

Next, we simply need to make use of this in our router. For example, let's secure the POST endpoint for our product API. Make it look like the following:

```
case '/api/products':
  if (req.method === 'GET') {
    // Find and return the product with the given id
    if (parsedUrl.query.id) {
      findProductById(id, req, res);
    }
    // There is no id specified, return all products
    else {
      findAllProducts(req, res);
    }
  }
  else if (req.method === 'POST') {
    var body = "";
```

```
req.on('data', function (dataChunk) {
  body += dataChunk;
});
req.on('end', function () {
  var postJSON = JSON.parse(body);

  // Verify access rights
  getTokenById(postJSON.token, function (err, token) {
    authenticator.tokenOwnerHasRole(token, 'PRODUCER',
    function (err, result) {
      if (result) {
        insertProduct(postJSON, req, res);
      } else {
      res.writeHead(403, {"Content-Type":
      "application/json"});
        res.end({
          error: "Authentication failure",
          message: "You do not have permission to perform that
          action"
        });
      }
    });
  });
});
}
break;
```

That's it! Implementation for the other endpoints is the same, and we will provide you with the full example source code for them.

Though I have covered some basics here, security remains one of the largest and most diverse areas of contemporary software development. We believe that token-based authentication will address a majority of the cases that you are bound to come across in your career. I would like to offer some suggestions for future study as well as complements to the topics that you have studied here.

OAuth

One of the most common authentication standards offered by modern web apps is **OAuth (Open Authentication Standard)**, its second version (**OAuth2**) in particular. OAuth makes heavy use of access tokens and is used by (among others) Facebook, Google, Twitter, Reddit, and StackOverflow. Part of what makes the standard powerful is that it allows users to sign in with their Google or Facebook accounts, or even some other account that supports OAuth2, when using your services.

There are several mature NPM packages for using OAuth2 with Node.js. In particular, we recommend you to study the node-oauth2-server package (`https://github.com/thomseddon/node-oauth2-server`).

Time-stamped access tokens

To keep things simple and focus on the main concepts, we have allowed our access tokens in this example to be permanent. This is a very bad security practice since tokens, like passwords, can be compromised and used to grant unauthorized users access to the system.

A common way to reduce this danger is to impose a **Time To Live** (TTL) value on each access token, indicating how long the token can be used until the user has to authenticate themselves again in order to get a new token.

Hashing passwords

For the sake of simplicity, we allowed passwords in this example to be stored and retrieved as plain text. Needless to say, this is an abysmal security practice and nothing that you should ever do on a production server. Mature Node.js frameworks such as Express.js provide built-in mechanisms for hashing passwords, and you should always choose those when available. In the event that you need to hash passwords on your own, choose the `bcrypt` module in order to both hash and compare. Here's an example of the same:

```
var bcrypt = require('bcrypt');

var userPlaintextPassword = "ISecretlyLoveUnicorns";
var userHashedPassword = "";

// First generate a salt value to hash the password with
bcrypt.genSalt(10, function(err, salt) {
  // Hash the password using the salt value
  bcrypt.hash(userPlaintextPassword, salt,
  function(err, hashedPassword) {
    // We now have a fully hashed password
    userHashedPassword = hashedPassword;
  });
});

// Use the same module to compare the hashed password with
potential
//matches.
```

```
bcrypt.compare("ISecretlyLoveUnicorns", userHashedPassword,
  function(err, result) {
    // Result will simply be true if hashing succeeded.
  });
bcrypt.compare("ISecretlyHateUnicorns", userHashedPassword,
  function(err,  result) {
    // result will be false if the comparison fails
});
```

Summary

In this chapter, you learned about token-based authentication and saw how it can work in practice to reinforce the backend. To put it into practice, we wrote a simple token-based access system to protect access to a set of backend data. Our server is now almost complete, but we must still deal with some other pressing concerns that modern apps need to face.

In the next chapter, we will explore how to address one of these most important concerns.

5
Real-Time Data and WebSockets

In this chapter, we will show you how to enable real-time data communication using WebSockets. This will allow your server to directly communicate with the connected clients without having any polling on the client side.

I would really love a two-way conversation, John

In its infancy, the Internet was not much of a two-way street. The traditional client-server architecture was the king, and servers initiating communication with clients was almost unheard of (and quite possibly seen as quite heretical by some, too).

However, starting with protocols such as **Internet Relay Chat** (**IRC**), real-time chat applications quickly became killer apps over time, with an enormous surge in popularity among ordinary users (if you are old enough to remember the **instant messaging computer program** (**ICQ**), yes, that is nostalgia that you are feeling). It was not very long until real-time features took the leap to the HTTP world, causing browser-based chat services to pop up everywhere. Meanwhile, related concepts such as push notifications gained popularity, especially with the advent of the smartphone.

Polling

Today, real-time features are an integral part of the Internet as we know it. However, their implementation has not always been optimal. Especially during its early stages, real-time data communication was almost always implemented by using polling, a technique where the client regularly contacts the server in order to check whether its state has changed. If it had (say, if a new message had been made available), the server responded by sending the updated state back. Needless to say, polling is a recipe for wasted resource usage. Moreover, it leads to rather choppy programming, since we find ourselves just repeatedly asking for updates rather than waiting and taking action on them when they are actually sent.

WebSockets

To avoid polling, we need a full-duplex solution, where the server can communicate directly with the client without the latter's initiative. Today, perhaps the most advanced and prevalent solution for this is the **WebSocket** protocol. A **WebSocket** protocol is a direct, two-way connection between the client and the server over the TCP protocol. It is structured in a way that allows both sides of the connection to initiate data transfer on their own. WebSockets were standardized in 2011 and are supported in all major browsers.

Using WebSockets in Node.js

Node.js does not come with a default module for using WebSockets. While we have tried so far to avoid third-party solutions and just show you how to work with **Vanilla node**, this topic is complex enough to put writing a WebSocket handler from scratch well beyond the scope of the book. Therefore, for the remainder of this chapter, we will use the excellent `socket.io` library. We of course do not imply that this is what you should be using in your own work, and in the *Chapter 6, Introducing Ionic*, the *Going further* section at the end of the chapter, we will direct you to alternative solutions and reading materials for WebSockets.

Setting up our project

We will set up a separate project for this chapter, demonstrating how we can create a simple chat application that demonstrates the essentials of using WebSocket.

Create a separate project folder and name it `chat-app`. In this folder, create a blank file named `app.js`. Finally, open your terminal or the command prompt, go into the folder, and run the following:

```
npm init
```

Answer the questions prompted by Node.js and make sure that you specify `app.js` as the entry point for the application.

Installing socket.io

We will install `socket.io`, as always, by using our good friend `npm`. From your terminal, issue the following command:

```
npm install socket.io
```

That's it. We are now good to go. Let's start setting up our server! However, before we do that, let's start from the top and define a basic chat interface for us to play with.

Creating a chat interface

We are not creating the next WhatsApp (yet!). So, building a full-fledged chat interface is a bit beyond what we want to achieve in this chapter. Let's go for something basic, as illustrated in the next screenshot:

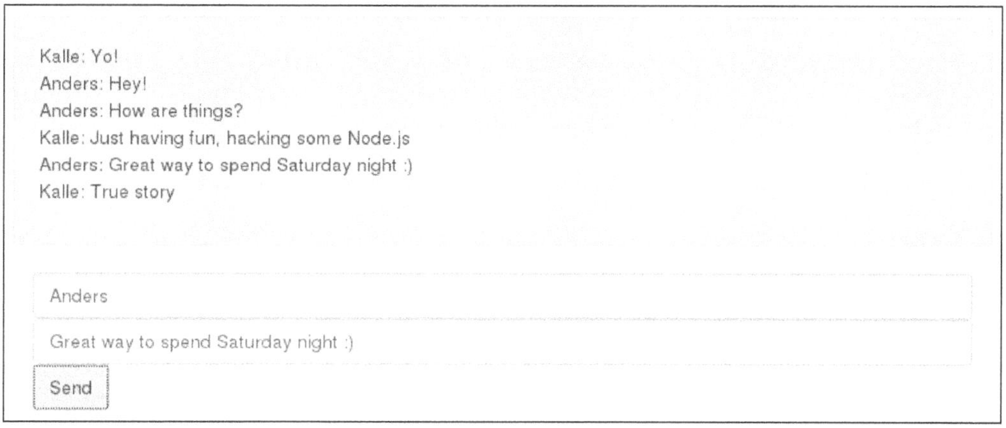

To create this layout, create the `index.html` file in your project folder and insert a basic HTML setup inside it, as follows:

```
<!DOCTYPE html>
<html>
  <head >
    <meta charset="UTF-8">
    <title>Socket.IO chat application</title>
  </head>
  <body>
  </body>
</html>
```

We will now add some custom elements to this markup in order to get the layout we need for our chat to be nice and user friendly. First, import the **Bootstrap** CSS framework by inserting a link into `href` in the header:

```
<head lang="en">
  <meta charset="UTF-8">
  <title>Socket.IO chat application</title>
  <link rel="stylesheet"
  href="http://maxcdn.bootstrapcdn.com/bootstrap/
  3.3.4/css/bootstrap.min.css"/>
</head>
```

Bootstrap, originally developed by Twitter, is a widely used framework that can be utilized to quickly build responsive web interfaces. As web design is beyond the scope of this book, we will use it in order to keep manual styling to a minimum. Don't worry if you are unfamiliar with the framework. It is very intuitive, and we will explain what you need to know along the way.

Next, let's add a Bootstrap container `div` to our interface, as follows:

```
<body>
  <div class="container"></div>
</body>
```

This is simply an organizational unit that Bootstrap uses to contain a set of UI elements inside a `container` so that the layout fits well on the screen being used.

Next, inside the `container`, let's add a `chat-box`, as follows:

```
<div class="row">
  <div id="chat-box" class="well">
    <ul id="chat-view" class="list-unstyled"></ul>
  </div>
</div>
```

The following are the three classes that are being used in the preceding code:

- The `row` class, which is similar to the `container` class, is an organizational unit that confines the elements that it holds to a single row in the layout.
- The `well` class, which creates a shaded container, make the elements it contains more visually distinct.
- The `list-unstyled` class, which simplifies the ordinary HTML unordered list tag, removes, among other things, the bullet styling that appears next to elements.

The end result is shown in the following screenshot:

Now, let's add the elements needed for users to enter their names and submit actual messages, as follows:

```
<form action="">
  <div class="row">
    <input type="text"
    id="chat-name"
    class="form-control"
    placeholder="Your name">
  </div>
  <div class="row">
    <input type="text"
    id="chat-message"
    class="form-control"
    placeholder="Enter message">
    <button id="chat-submit"
    type="submit"
    class="btn btn-default">Send
    </button>
  </div>
</form>
```

By now, you should be familiar with most of the UI elements and what they do, and the rest is nothing but a standard HTML form (note that we do not provide an action for the form itself; submissions will be handled dynamically through JavaScript instead). Note that we added some classes to the `form` elements. These are standard Bootstrap layout classes that are used to style the appearance of the elements themselves. They do not introduce any functionality in themselves, and as such, we do not need to deal with them in detail here.

That's it! If you open the file in your browser, you will see the following:

The chat obviously does not really do anything at the present time. We will do something about this in a moment, but first, let's see how we can serve the HTML file that we just created directly from Node.js.

A basic file server

We now have an HTML file for our interface, which we would like the user to see whenever they connect to the app via their browser. To make this happen, we need to make our Node.js app listen to HTTP requests and then respond with the appropriate HTML file. Sounds familiar? Yep, it's time to reintroduce the Node.js HTTP module. Go ahead and add the following at the top of the app.js file:

```
var http = require('http');
var url = require('url');
var fs = require('fs');
```

We have already seen the first two modules. The third one, fs, is the standard module that is used to handle interactions with the file system. We will need this module in order to retrieve and serve the HTML file.

Let's create an HTTP server for this end. Add the following to app.js:

```
var server = http.createServer(function (req, res) {
  var parsedUrl = url.parse(req.url, true);
  switch (parsedUrl.pathname) {
    case '/':
    // Read the file into memory and push it to the client
    fs.readFile('index.html', function (err, content) {
      if (err) {
        res.writeHead(500);
        res.end();
      }
      else {
```

```
            res.writeHead(200, {'Content-Type': 'text/html'});
            res.end(content, 'utf-8');
        }
    });
    break;
    }
});
```

Let's go through what happens here. Upon receiving an HTTP request, our server will try to find a match for the path name of the request. If the path is for the root of the document hierarchy (signified by a normal slash), we want to serve the index. html document. If this is the requested path, the following happens:

1. The readFile() method, which is part of the fs module, is invoked in order to load the index.html file.

2. If the load fails (that is, if there was an I/O error), the server responds with status *500*, indicating a server error.

3. If the file is successfully loaded, we add its content (in this case, a string of HTML content) to the response payload, set the appropriate media type and code for the response, and serve it back to the client.

By default, a status *404* is served if the client tries to access any other part of the document hierarchy.

Let's see this in action. Add the following to the end of app.js:

```
server.listen(8080);
```

Start the server from your terminal, as follows:

node app.js

Open your browser and visit http://localhost:8080. You will see the following:

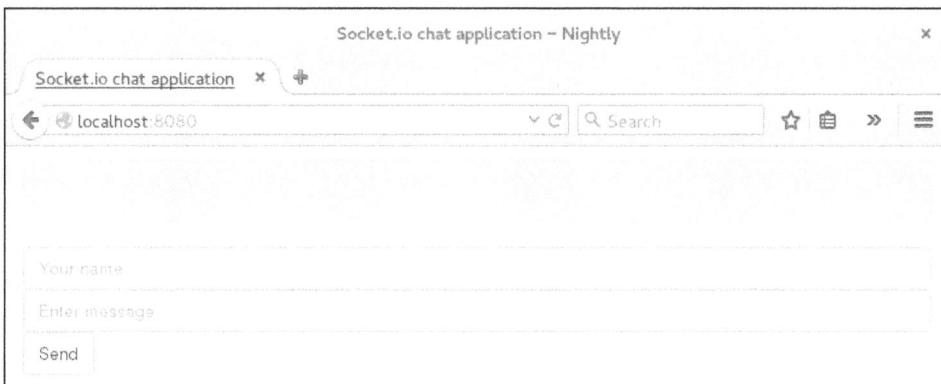

An alternate chat

You can find an excellent tutorial on how to build a chat application that is similar to ours using Express.js and `Socket.io` at `http://socket.io/get-started/chat/`.

In fact, the example that we used here is inspired by this one, though we have modified it in order to make it suitable for the book.

The efficient serving of static files

To keep things brief, and also to get our feet wet when it comes to using the HTTP module, we served a static HTML file directly from Node.js in our example. In a production system, static files are usually much more efficient as regards serving from a standard file server such as Apache or Nginx.

Summary

In this chapter, we covered the basic theory behind WebSockets and why they matter to modern, real-time web applications. Building on this, we created a simple chat application by making use of the `socket.io` library in order to demonstrate real-time communication between several clients connected to the same server.

6
Introducing Ionic

Ionic is a free and open source library of mobile-optimized HTML, CSS, and JavaScript components, gestures, and tools that can be used to build highly interactive mobile apps. Being built around the widely used **Syntactically Awesome Stylesheets (Sass)** and AngularJS technologies, Ionic offers web developers with a basic knowledge of HTML, CSS, and JavaScript an opportunity to develop cross-platform mobile applications.

In this chapter, we'll cover the following:

- Setting up your Ionic web services
- Creating your first Ionic application
- Using Ionic View to test your applications
- Sharing your creation with your collaborators

Setting up your Ionic web account

In *Chapter 1, Setting Up Your Workspace*, we highlighted the basic essentials of setting up your workspace for Ionic, which included an installation of the core Ionic libraries as well as the Android and iOS SDKs.

In this section, we will further elaborate on setting up an account on `ionic.io`, which is the web service that allows us to easily deploy and test our creations on Android and iOS devices through the Ionic View application. Through the Ionic web service, we will also be able to configure capabilities such as push notifications, which are an excellent way of interacting with your application's audience.

In order to start off with setting up your `ionic.io` web account, visit `apps.ionic.io` and click on **Sign Up**:

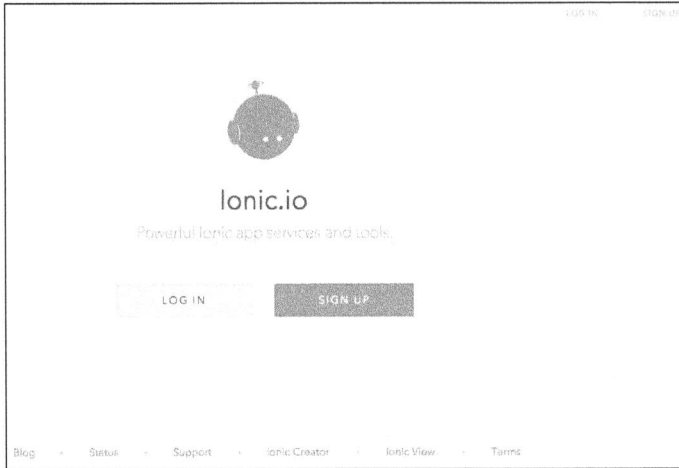

The signup screen looks like the following screenshot:

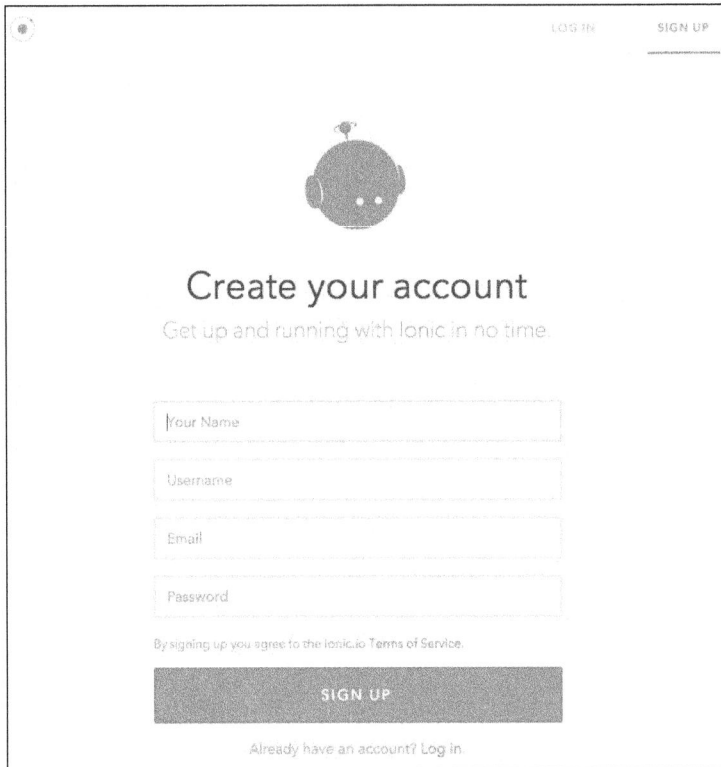

When filling in the essential details, you'll be welcomed with the following screen, which will give you an overall view of setting up your first Ionic project:

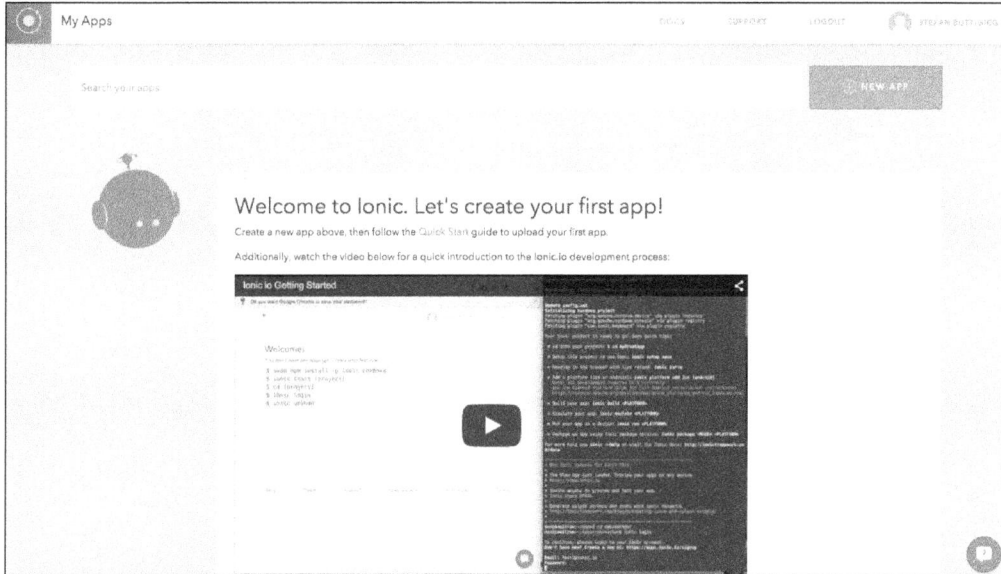

Creating your first Ionic application

To start off with your first Ionic project, open up your terminal and run the `start` command, as follows:

```
$ ionic start myfirstionicapp
```

Then, you will need to change the directory to your Ionic project directory, which is the same as your project title:

```
$ cd myfirstionicapp
```

Once you've navigated to the right directory, you will need to log in to your Ionic web account with the following command. This will be followed by inputting the email address associated with your account and your password:

```
$ ionic login
```

Once your credentials are verified, you will be able to upload your first creation to the Ionic web service with the following command:

```
$ ionic upload
```

Once you upload your application, you will be able to see your application with the Ionic web service apps dashboard, where you will be able to see all your Ionic apps:

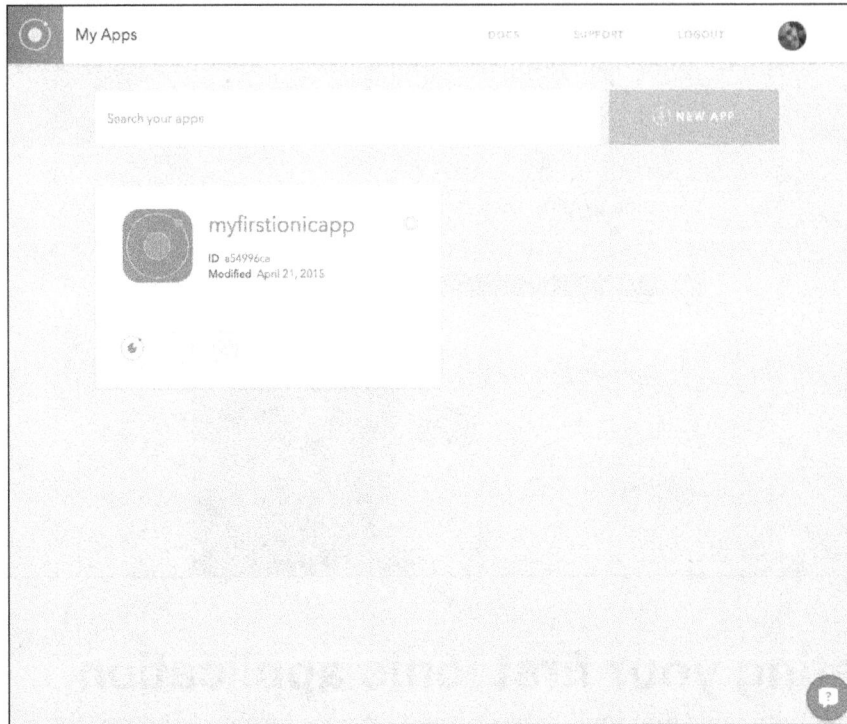

Checking out your Ionic applications with Ionic View

Ionic has launched a very nifty application for iOS and Android where you will be able to see and test your application creations on your smartphone within minutes. We will first need to start off by installing Ionic View on your particular device.

Installing Ionic View on Android

Search for the Ionic View App on the Google Play Store and download it to your device.

When you load the application for the first time, you'll see the following screen, where you will be asked to enter your login credentials:

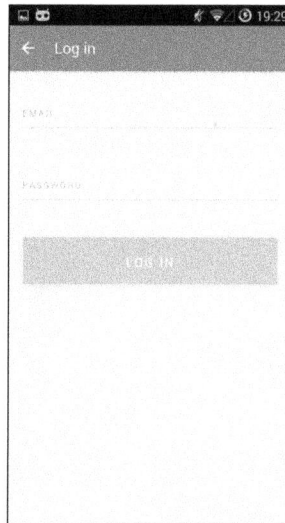

Once you sign in, a dashboard with your current application should show up.

Installing Ionic View on iOS

Search for Ionic View on the App Store and download it to your device. Once you install the application, you'll see the following, where you will be asked to enter your login credentials:

This is how the **LOG IN** page looks like:

When you log in for the first time, depending on whether you managed to successfully upload applications to the Ionic Apps web service by following the previous instructions, you should be able to see something like this:

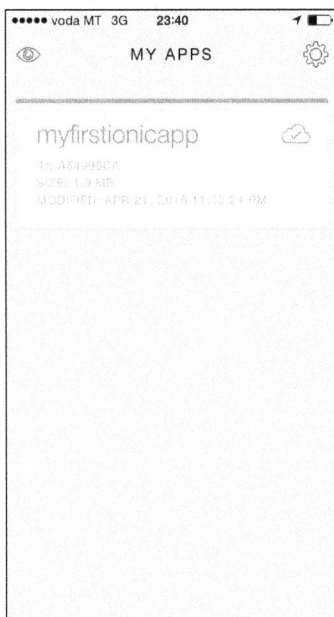

Testing your application on the iOS Ionic View App

Once you have loaded the application for the first time and logged in, you will be able to start testing your applications. You can proceed with this stage by downloading your application to your device by tapping on **Download App**:

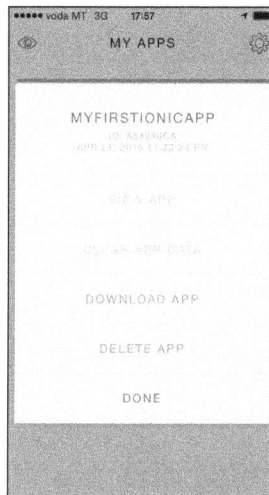

After downloading the application, you will see what's shown in the following screenshot, where you will have an option to view the application on your device. In order to try out your application, tap on **View App**:

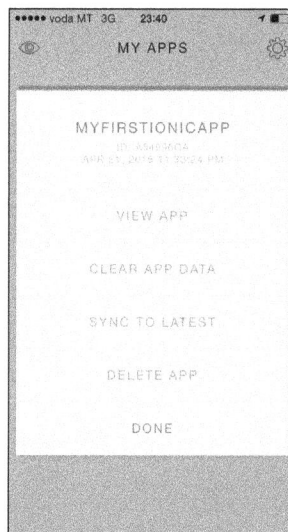

Your first Ionic application is based on a boilerplate provided by the Ionic framework to highlight the capabilities of Ionic. To exit the application once you have finished testing the application, you need to swipe down on the screen with three fingers:

Testing your application on Android

After you log in from the Ionic View App, you will see what's shown in the following screenshot, where you are provided with a dashboard of all your Ionic applications. To start testing your applications on Android, use an approach similar to that for the iOS Ionic View App, where you need to download your app to the device:

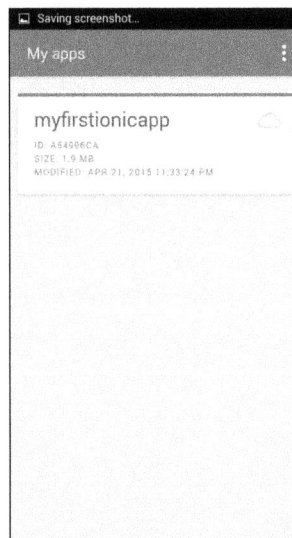

After tapping on **myfirstionicapp**, tap on **Download files** to synchronize the application data from the Ionic web service to your device:

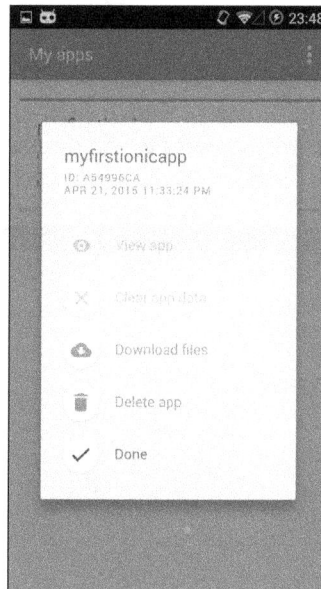

Once the application is synced to your device, you will see the following. This enables **View App**. Tap on the corresponding button to view your application:

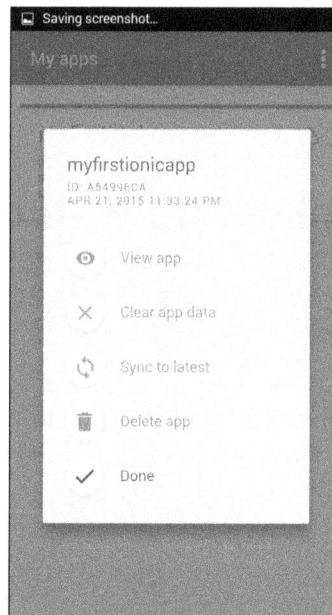

Once you load the application, you will see the following application, which is fully enabled. This application works within the Ionic **View App** with the impression of a working application. As previously note, the application that we created together is based on a boilerplate produced by the Ionic framework team.

As you may have observed, the design conventions followed for this mobile application are suitable for each platform, which further enhances the value proposition that the Ionic framework offers to the developer. That is, cross-platform applications are adapted to the platform that the device is based on:

Sharing your application with collaborators

Sharing your application with your collaborator is quite straightforward. This is possible via the Web Service and command line. Doing it via the web service requires you to click on the gear icon on the Ionic app of your choosing and then click on **Share**.

From the command line, you'll first need to navigate to your project directory via the terminal. Then, you will need to log in by typing the following command:

```
$ ionic login
```

Once you have logged in, you just need to type the following command and replace EMAIL with the e-mail address of your chosen collaborator:

```
$ ionic share EMAIL
```

Your collaborator will receive an email with an invitation to view the app. We recommend that they view this email from a mobile device to be able to see the app.

Going further

There are a number of incredible features within the Ionic framework's ecosystem, which take the framework to the next level. It's worth exploring the different features of the Ionic framework by checking out their documentation, which is available at http://docs.ionic.io/.

If you would like to experiment further, it will be worthwhile at this stage to explore how to set up push notifications for your Ionic app projects, Ionic lab, Ionic analytics and deploy.

Ionic lab allows you to view the Android and iOS version of your Ionic app side by side in the browser, which gives you an opportunity to see the salient differences between the different platforms that are supported by Ionic. In the next chapter, we will use Ionic lab more frequently, as you should now be able to understand how to build different user interfaces for our different needs.

Summary

In this chapter, we covered all the basic essentials of setting up a working environment to efficiently create and share Ionic apps. As previously mentioned, there are a number of other tools, which we haven't covered in detail in this chapter, that are worth trying out and which will help you effectively create your app's workflow.

In the next chapter, we will learn the general structure of an Ionic project and how its components work together to create mobile experiences. These components are based on web technologies and yet look, feel, and work like native applications.

7
Building User Interfaces

In this chapter, you will learn how to add new and unique elements to and modify your current Ionic project from a practical point of view. Among other things, I will show you how to modify the tab icons and add a new tab controller. This chapter will guide you and give you the necessary knowledge that is required to have a deep understanding of how to create and modify your own mobile applications.

Most importantly, you will also get acquainted with the AngularJS JavaScript framework, which lies at the heart of the functionality of Ionic.

The entire source code of this modified project is available on the GitHub repository, which can be viewed by visiting `https://github.com/stefanbuttigieg/nodejs-ionic-mongodb`.

The structure of an Ionic project

In the process of creating a project in the previous chapters, we created a directory entitled `myfirstionicapp`, which can be found in the root folder. We recommend that you open this project folder with an **Integrated Development Environment (IDE)** or a text editor. In our case, we are comfortable using Atom, which is an advanced open source text editor.

> You can download Atom from `https://atom.io/`.

Once you open your IDE and add the project folder to your workspace, you will see the following folder structure:

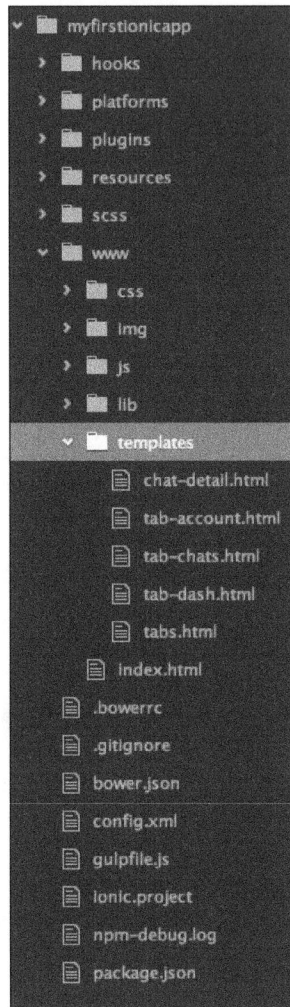

```
myfirstionicapp
  > hooks
  > platforms
  > plugins
  > resources
  > scss
  www
    > css
    > img
    > js
    > lib
    templates
        chat-detail.html
        tab-account.html
        tab-chats.html
        tab-dash.html
        tabs.html
      index.html
  .bowerrc
  .gitignore
  bower.json
  config.xml
  gulpfile.js
  ionic.project
  npm-debug.log
  package.json
```

Let's take a closer look at each of the folders shown in the preceding screenshot in turn:

- `hooks`: This folder is where our project will store automatically (or manually) generated hooks for the underlying Cordova system, which provides most of of our project's runtime.

- `platforms`: This folder contains the necessary files and configurations that are required to deploy a project on a specific platform, such as Android or iOS.

- `plugins`: This stores the various Cordova plugins for our project. If you examine it closely, you will observe that it already contains a number of default plugins such as `com.ionic.keyboard`, which makes it easier to work with each platform's native keyboard functionality.

- `resources`: This contains global and platform-specific resources, such as app logos, splash screens, and so on.

- `scss`: This contains the core **Sass (Syntactically Awesome Style Sheets)** files for our projects. By modifying these, we can extensively alter the appearance of our app on the various platforms that it targets.

- `www`: This is the folder that you will find yourself working with more than any other folder.

  If you have previously worked with web apps, its contents will be familiar to you:

  - `css`: This contains the CSS files for your app.

  - `img`: This contains the static images for your app.

  - `js`: This contains the JavaScript files for your app. This is also where most of your custom application logic will reside.

  - `lib`: This contains third-party libraries and applications that can be used in your project. Among other things, this folder contains AngularJS itself, along with its associated dependencies.

  - `templates`: This contains the AngularJS template files, which are HTML files that may contain AngularJS-specific content, such as data bindings and directives (don't worry, we will explain what these are in a bit).

As mentioned earlier, your own work will predominantly be confined to the `www` folder. This makes sense, since the projects that we create with Ionic are actually a special breed of web apps that are customized to run on mobile devices.

Now that we are familiar with the structure of our project, let's dip our toes into AngularJS, the framework that makes it all work. It only gets better.

Introducing AngularJS

Ionic is powered by the AngularJS framework (which is also commonly just called Angular), which drives the UI interactions, gestures, animations, and well, essentially the entire functionality of your app. Understanding it is crucial to the experience of working with Ionic.

Angular was initially developed by Google in 2009 in an effort to enhance HTML with dynamic data binding at the tag level (the name **Angular** refers to the **angular brackets** around the HTML tags). Its architectural philosophy is firmly grounded in the **Model-View-Controller** (**MVC**) pattern and centered around an augmented HTML syntax for building UIs and a feature-rich, modular core framework to create business logic.

Due to its extensive nature, writing a concise introduction to Angular is not easy. As we work our way through the coming chapters, we will gradually go deeper and increase our knowledge of the framework. Here, we will settle for an outline of the most important aspects of AngularJS so that you can understand how these aspects work in the context of Ionic.

The structure of an Angular app

As we work our way through this chapter and the ones that follow, you will very quickly realize that what you are building with Ionic are actually augmented Angular apps that are designed for mobile devices. Since this is the case, it is crucial that you understand how Angular apps are structured.

Modules

The most fundamental module of an Angular app is, well, the module. A **module** is a collection of **services**, **controllers**, and **directives**, which provide some specific functionality to your app. In fact, your Angular app is itself a module!

Defining a module is rather simple:

```
angular.module('starter', []);
```

This creates a `module` named `starter`. The second argument is meant to contain a list of dependencies (more on this will be discussed later). This argument is left empty if the module does not depend on any other modules.

Modules within modules within modules

Modules can load other modules, incorporating their functionality into their own. This makes it very easy for developers to write and share utility modules, which can be used by other developers in their own apps (at the time of writing this book, there are literally tens of thousands of such modules hosted on GitHub, with many under active development).

Remember those empty brackets in the example that we saw just a bit earlier? This is where you list all the modules the current module should load for its own use. For example, in our Ionic apps, the `ionic` module is a fundamental component that we always want with us:

```
angular.module('myapp', ['ionic']);
```

Now, whenever this module is loaded, Angular will automatically load its dependencies with it.

Services, controllers, and other beasts

As mentioned before, the Angular modules contain other components, which provide various kinds of functionality to the app. Detailing them here would just clutter things, so we will introduce them as we go along (if not here, then in the later chapters, where they are needed). For now, it is sufficient that you just know that they exist and they together make up the functionality of an Angular module.

The Angular MVC pattern

Now that we have a better understanding of how an Angular app is structured, it is time to look at how it actually works during runtime.

The functionality of an Angular app revolves around the following three core concepts:

1. The **view** is what the user sees and the medium through which the user primarily interacts with and reads output from your application.

2. The **controller** responds to the user interaction with the application and communicates with the model in order to produce appropriate data. It then updates the view to reflect that data.

3. The **model** is a collection of data, libraries, services, and other things that make up your application's business logic. The model is responsible for the heavy processing in your app, and it is usually where most of your code will reside.

These three concepts make up the MVC pattern—model-view-controller. This is a very popular design pattern for modern web apps.

Now that we know how an Angular app functions, let's see how it realizes each of these three concepts.

The view

In an Angular app, the view is composed predominantly of standard HTML, which is augmented by Angular-specific components in order to facilitate dynamic updates. The following are the two primary components:

- **Directives**: These are the custom HTML tags, whose function and behavior are defined from within AngularJS but written like plain HTML. For example, a tag like the following can be a directive that draws a map centered on a specific latitude and longitude:

```
<map lat="39.234" lng="43.453"></map>
```

- **Expressions**: These are the expressions that are surrounded by double curly braces, which evaluate to a given value during the runtime of the application. Unless specified otherwise, the output of an expression will be updated as soon as the model of the application changes. The following is an example of such an expression:

```
{{ person.firstname }}
```

  The preceding expression does something that is very common in Angular – resolve the value of some object's member. However, to do so, we first need to define where that object can be found. This is where controllers come into the picture.

The controller

In an Angular app, the **controller** is realized by special module components, which are fittingly called controllers. You can define them in a module in the following way:

```
angular.module('myapp.controllers', [])
controller('MyCtrl', function($scope) {})
```

The first parameter is the name of the `controller`. The second parameter is a `function` that defines what the controller actually does. This `function` can take a variable number of arguments, which represent the dependencies that the controller will use, much like the way we defined dependencies for modules earlier.

The model

Broadly speaking, the **model** is *everything else* in your app. It is the sum total of the data models. Throughout the following chapters, we will gradually explore the various components that you can use to compose your model.

Putting it all together

Let's finish our brief tour of Angular by showing how to connect the various components that we have seen so far.

Consider a situation where you first navigate to the `index.html`, which is available at the following path `myfirstionicapp/www/index.html`:

When you navigate here, you will observe the following block of code:

```
<ion-nav-bar class="bar-stable">
  <ion-nav-back-button>
  </ion-nav-back-button>
</ion-nav-bar>
```

This block of code determines the header bar of the application, and this is one of the examples of the **User Interface** (**UI**) components, which can be managed through HTML5.

For documentation and reference purposes, you can refer to the Ionic UI components at `http://ionicframework.com/docs/components`.

As you further explore your project, you will see that the main controllers that will power the interactive functionality of your project are available at the following path:

myfirstionicapp | js | controllers.js

Modifying an Ionic project

In order to build upon the knowledge that we have gained and the work that we have previously done, we will modify the user interface of the project that we previously created. We will start off by modifying the header.

Modifying the header

Let's say that we would like to change the header bar to a calm blue color. Navigate to the `index.html` file available at **www | index.html**.

Refer to the `body` block, and using the reference UI components, change the `ion-nav-bar` class to the following:

```
<ion-nav-bar class="bar-positive">
  <ion-nav-back-button>
  </ion-nav-back-button>
</ion-nav-bar>
```

Modifying the tab colour, icons, and names

Since we have decided to change the header color, we will go ahead and modify the tab bar to make its color match the header color. We should first navigate to the `tabs.html`, file which is available at **www | templates | tabs.html**, and change the `ion-tabs` class to the following:

```
<ion-tabs class="tabs-striped tabs-icon-top
tabs-background-positive tabs-color-active-positive">
```

The `icons` need to be further modified to contrast with the new `blue` color. So, we will further modify the `ion-tabs` class to the following:

```
<ion-tabs class=" tabs-striped tabs-icon-top
tabs-background-positive tabs-color-light">
```

We will take a step further and attempt to change the icon's graphic. Let's say that we would like to change the dashboard icon to something that looks more circular.

First of all, we need to refer to the Ionicons documentation, which is available at `http://ionicons.com/cheatsheet.html`, and find out the class name in relation to the circular analytics icon. For this example, we will use `ion-ios-analytics`. When we want the user to tap on the icon and activate the dashboard, we want the icon to be highlighted, whereas when it's not active, we need the user to see an outline of the icon. In order to achieve this, we will need to declare the icons that will be used in both an active and inactive state.

In order to do this, we will navigate to the `tabs.html` file and modify the `Dashboard` tab icon in the following way:

```
<ion-tab title="Dashboard" icon-off="ion-ios-analytics-outline"
icon-on="ion-ios-analytics" href="#/tab/dash">
  <ion-nav-view name="tab-dash"></ion-nav-view>
</ion-tab>
```

Modifying our pages

In this particular example, we will edit the dashboard page, where we will modify the content of the **list card** UI components.

The modification of the dashboard is possible by navigating to the `tab-dash.html` file, where we will see the different cards declared in the `div` class of the `list card`.

The `list card` is declared as follows:

```
<div class="list card">
  <div class="item item-divider">Title of List Card</div>
```

```
      <div class="item item-body">
      <div>
      List Card Content
    </div>
  </div>
```

By using the Ionic framework, it's possible to include a footer to your card. In our case, we will add a footer to the `Health` list card, declaring that the user has walked `10,000` steps today. In order to do this, we will add an `item-divider` class, thus modifying the `list card` as follows:

```
<div class="list card">
  <div class="item item-divider">Health</div>
  <div class="item item-body">
    <div>
      I really can!
    </div>
  </div>
  <div class="item item-divider">
    Great Job, You did 10,000 steps today!
  </div>
</div>
```

All the modifications that you made until now will result in a **Dashboard** tab, which will look like the following screenshot:

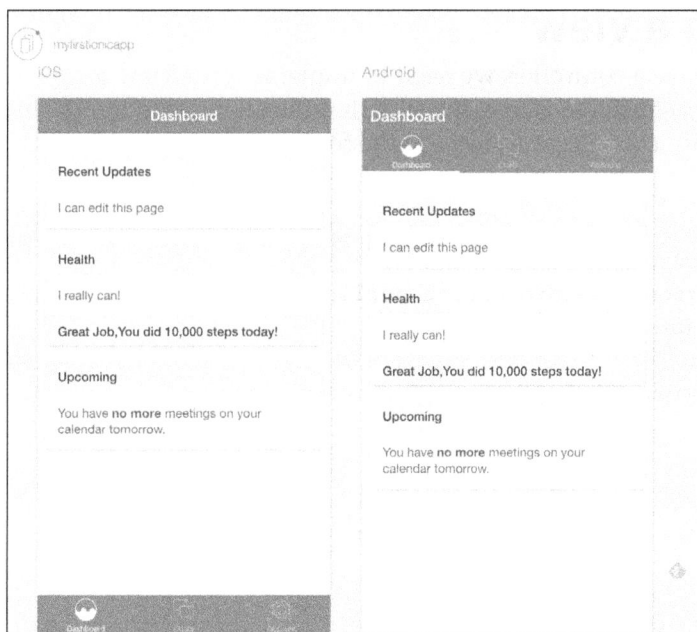

Adding a new tab

In this project, we would like to add a new tab that highlights the developers of this application. This will inform the potential users of this app about how to get in touch with the developers to provide them with the necessary feedback to improve the app experience.

In order to achieve this, we'll need to perform the following four main steps:

1. Create a new controller entry in `controllers.js`.
2. Create a new tab controller called `tab-about` in HTML.
3. Add a new tab entry in the `tabs.html` file.
4. Consolidate our work in the `app.js` file and connect everything together.

Creating a new controller

Let's start with adding a controller for the new tab. Head over to `controllers.js` and add the following into it:

```
controller('AboutCtrl', function($scope) {})
```

Don't worry about the empty function for now. At the moment, our tab does not need any functionality apart from simply appearing.

Creating a view

Now that we have a controller, we need to implement the view for the new tab. The first order of business is to make sure that the tab is added to the list of tabs. To do so, modify `tabs.html` to include the following:

```
<!-- About Tab -->
<ion-tab
title="About"
icon-off="ion-ios-information-outline" icon-on="ion-ios-information"
href="#/tab/about">

<ion-nav-view name="tab-about"></ion-nav-view>

</ion-tab>
```

This creates the fundamental bindings for the new tab and adds it to the list of tabs. However, we will still need to add the content that should open when the user clicks on the tab. To do so, create a new file called `tab-about.html` in the `templates` folder and put the following code in it:

```
<ion-view view-title="About">
  <ion-content>
    <div class="list card">
      <a href="#" class="item item-icon-left">
        <i class="icon ion-ios-people"></i>
        Christopher Svanefalk and Stefan Buttigieg
      </a>
      <a href="#" class="item item-icon-left">
        <i class="icon ion-home"></i>
        Malta and Sweden
      </a>

      <a href="#" class="item item-icon-left">
        <i class="icon ion-ios-telephone"></i>
        +3569912345678
      </a>
      <a href="#" class="item item-icon-left">
        <i class="icon ion-ios-world-outline"></i>
        www.ionicframework.com
      </a>
    </div>
  </ion-content>
</ion-view>
```

Adding a state for the new tab

Next, we need to add a new navigation state to the `controller` in order to allow the user to navigate, with the help of clicks, to the `tab-about.html` tabs content page. To do so, open the `app.js` file and add the following state:

```
state('tab.about', {
  url: '/about',
  views: {
    'tab-about': {
      templateUrl: 'templates/tab-about.html',
      controller: 'AboutCtrl'
    }
  }
})
```

Note that the following is what the preceding code does:

- The `url` property determines whether the application enters into the state of accessing the `/about` URL.

- Inside the `views` property we determine the path to the `view`, which should be loaded when this application enters into the `view` state. In this case, it is the `tab-about.html` file that we created earlier.

- Finally, inside `views`, we also determine which `controller` is responsible for handling this application state. In our case, it is the `AboutCtrl` controller, which was defined by us earlier.

Testing the newly created tab

Quick testing is possible through your local browser. Once you save your project files with your IDE, you will be able to see your app in the prototype form through your browser:

1. First navigate to the project folder:

   cd myfirstionicapp

2. Then, type in the follow command:

   ```
   ionic serve --lab
   ```

The results for this are shown in the following screenshot. These results are adapted for both iOS and Android. In addition to this, you'll be able to test your application through a point-and-click interface. This experience is similar to having an iOS or Android emulator working through a browser:

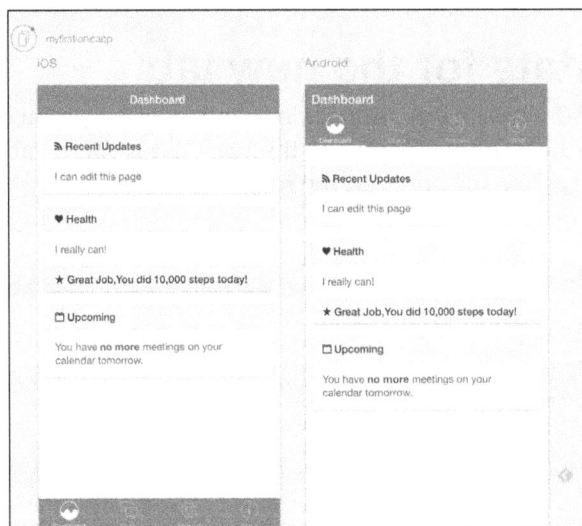

Going further

The importance of setting out your layout and user experience before you start off with any project is crucial. This will enhance your user experience, and it will be even more helpful not only while laying out the necessary project structure, but also throughout the programming process.

You can further customize the look and feel of your application by using Sass. Sass is marketed as being the most mature, stable and powerful professional **grade CSS extension language**, and it allows you to further customize your project.

In order to improve your knowledge of AngularJS, we recommend that you check out a free video resource, which is available at `http://campus.codeschool.com/courses/shaping-up-with-angular-js`, `http://campus.codeschool.com/courses/shaping-up-with-angular-js`.

The aforementioned course is sponsored by Google and is available for free for anyone who would like to dive deeper into AngularJS. Experimenting with different Ionic user components has become easier than ever, especially with the recent **Playground** tools that are available in **Ionic Playground**, which is available at `http://play.ionic.io`.

From the Android point of view, in order to take your project to the next level, there is a free library available, which can be used to integrate the latest iteration of Google's interactive Material design by installing the Ionic Material library. This can easily be installed by first ensuring that you have bower installed and then navigating to your project folder and inputting the following command:

```
bower install ionic-material
```

We are hopeful that with this chapter, you will start experimenting with your very own projects while staying in line with providing remarkable user experience with the current knowledge that you've gained until now. In the next chapter, we'll take a step further.

Summary

In this chapter, we took the template project that we created in the past few chapters and modified it to something closer to what we envisioned by modifying the appearance of the user interface. In addition to this, we also started experimenting with the Ionic project code to better understand what the project is made up of.

Furthermore, we provided you with the basic knowledge to start prototyping your Ionic application within the browser.

8
Making Our App Interactive

In the previous chapter, we gave you a gist of how to work with user interfaces in Ionic. While doing so, we also gave you a thorough introduction to AngularJS, the driving force behind Ionic apps, and explained how you can use it in order to let users interact with your app.

Here, we will continue building on this while simultaneously exploring new features of Ionic, including the interaction with the native features of your device. We will put this all into practice by building a concrete app that will load and display the contacts from your phonebook. By doing so, you will also learn how to compile and run the Ionic apps on physical devices (such as iPhones and Android phones), rather than the emulator that we have used so far.

Creating a new project

Ionic comes with a powerful **Command Line Interface** (**CLI**), which quickly lets you create, modify, and extend Ionic projects. By automating chores such as module integration and scaffolding, it can drastically improve your productivity.

'Let's use the CLI to create a basic project for us to work with in this chapter. Go to your working directory and run the following in a terminal:

```
ionic start phonebook-app blank
```

This will create a blank app containing only the basic components for a bare-bones working app. This is suitable for us, since we want to build an app from the ground up and learn as we go.

> The `ionic start` command has several other basic templates as well. For example, the `tabs` template gives you a basic app with tabbed navigation that you can extend, while side menu creates a basic app with a side menu for navigation.

When this command finishes running, you will have a complete project to work with. No extra fiddling is required! Now, let's go ahead and add some content in it.

Creating a basic MVC project

Our goal in this chapter is to create an app that can pull information from the contacts storage of the local device and display it to the user (a phonebook, if you like). To do so, we need to do the following things:

- Define a **view** (template file) to display the contact list
- Define a **controller** in order to handle interactions with the list
- Provide the necessary **model** logic in order to provide contacts' information.

You may recall that this workflow fits nicely with the overall architecture of AngularJS, which follows the MVC pattern. We will take care of each item in turn.

Creating the view

Go ahead and add the following folder to your projects:

    www/templates

Here, we will store all the view templates that we will use throughout our project. In this folder, let's create our first view file, as follows:

    www/templates/contacts.html

If we need to add additional views to our app, we will do it in the same manner. Partitioning our views like this not only makes organization easier, but also boosts performance, since HTML will only be loaded on demand when it is needed, rather than all at once.

Let's add some content in order to create an actual view:

```
<ion-view view-title="contacts">
  <ion-content>
    <ion-list>
```

```
        <ion-item>
        </ion-item>
      </ion-list>
    </ion-content>
  </ion-view>
```

Don't concern yourself with the list-related tags yet; we will get to what they do in a moment. For now, let's look at the two outer ones:

- `ion-view`: This tag tells Ionic that this is a view that can be dynamically loaded from other parts of the application. We add the view-title attribute to it in order to create a label that can be used to refer to the view.

- `ion-content`: This tag designates a content area in the view, which is especially good at displaying scrolling data. Since we want to display a list of contacts of an unknown length, this is what we will want to wrap our list in.

Creating the list view

Lists are some of the most ubiquitous data structures in apps everywhere. So, it is no surprise that most frameworks provide powerful tools to work with them. Ionic is certainly no exception.

Have a look at the code that we added earlier. Especially note the tags inside the `ion-content` tag:

```
<ion-view view-title="contacts">
  <ion-content>
    <ion-list>
      <ion-item>
      </ion-item>
    </ion-list>
  </ion-content>
</ion-view>
```

The following two tags encapsulate the majority of Ionic's list rendering capabilities and are generally everything that you need in order to display the list:

- `ion-list`: This indicates that the wrapped content is a list
- `ion-item`: This indicates a single data node, which can be rendered in a list

Since each item in our list will be a single contact, we will wrap each of them in an `ion-item` tag.

At present, our view does not really do much apart from displaying an empty list. We need to add some markup in order to show the details of the contact wrapped in each item. Let's do something basic first, such as just showing the name and mobile number of the contact, if any:

```
<ion-view view-title="contacts">
  <ion-content>
    <ion-list>
      <ion-item>
        <h2>{{contact.name}}</h2>
        <p>{{contact.number}}</p>
      </ion-item>
    </ion-list>
  </ion-content>
</ion-view>
```

Here, we defined two Angular expressions to render the name and number of a contact during runtime. The contact in this case is simply a **JavaScript Object Notation (JSON)** object holding information about a given person in our contact list. Next, we will see how to assign a concrete value to this JSON object; this will take place in the associated controller.

> As of AngularJS 1.3, which is the version that is officially shipped with Ionic at the time of writing this book, it is possible to make expressions behave in a bind-once fashion. This means that they take on the values that they initially compute to, and the values are skipped by the AngularJS DOM update cycle after this, which improves performance. In our case, this works well, since we will only display our data once after fetching it, as we will see later. The downside of this method at this point in time is that updates to the data model will not be reflected in this view.

To make your expressions bind-once, add two colons (::) before them, as follows:

```
<ion-view view-title="contacts">
  <ion-content>
    <ion-list>
      <ion-item>
        <h2>{{::contact.name}}</h2>
        <p>{{::contact.number}}</p>
      </ion-item>
```

```
        </ion-list>
      </ion-content>
    </ion-view>
```

Creating the controller

You may recall that the controller is the *glue* between your view and model (that is, your business logic). Its primary responsibility is to handle interactions with your UI from the user and delegate the processing of an appropriate response to the model.

In the previous chapter, the example controllers were located in a separate JavaScript file, where they were declared as part of a module. This is a good design practice, since it makes it much easier to structure your app.

Note that our blank project does not have such a file. So, let's start by creating it. Add the following file to your project:

```
www/js/controllers.js
```

In this file, let's add the following:

```
angular.module('phonebook.controllers', [])
controller('ContactsCtrl', function ($scope) {
});
```

Let's recap in brief to see what the preceding code does:

1. We defined a module using the core `angular.module()` function, which takes the following parameters:

    ○   The first parameter is `phonebook.controllers`, which is the name of the module.

    ○   The second parameter is an empty list, which indicates that this module has no external dependencies. If it did, we would list their names here so that the AngularJS dependency injection system can resolve them during runtime.

2. Having created a module, we attach a controller to it using the (aptly named) core function `controller()`. The following arguments are passed to this function:

    ○   `ContactsCtrl`, which is the name of the controller, is the first argument.

○ The second argument is a `function` that defines what the controller does. In this controller, we will define any and all the actions that are taking place in the segment of the app controlled by this controller. Note that this function takes the `$scope` parameter (we will get to what it does a bit later). Just like the dependency list in the module, this parameter and all the others passed to the function (there can be any number of parameters or none at all) denote a dependency of this controller, which will be resolved at runtime using the dependency injection system.

For now, this is all we need in order to have a full-fledged controller. Next, we will need to connect it with our view.

Connecting the view and controller

We have the following two choices if we need to bring our newly created view and controller together:

- We can use an inline view, where we put a reference to the view and controller directly in the `index.html` file or another template, which is in turn loaded from `index.html`
- We can use a router to associate the view and its controller with a certain path within the application

Even though we only have one view for now, we will go for the second option. This might seem redundant, but it makes it much easier to structure the app in the event that we want to add new navigation states later on (spoiler — we will!)

Routing is normally configured in the `app.js` file. So, that is where we will go next. Open the file and make sure that it has the following content:

```
angular.module('phonebook', ['ionic', 'phonebook.controllers'])
.run(function ($ionicPlatform) {
  $ionicPlatform.ready(function () {
    if (window.cordova&&window.cordova.plugins.Keyboard) {
      cordova.plugins.Keyboard.hideKeyboardAccessoryBar(true);
    }
    if (window.StatusBar) {
      StatusBar.styleDefault();
    }
  });
})
.config(function ($stateProvider, $urlRouterProvider) {
  $stateProvider
```

```
  .state('contacts', {
    url: '/',
    templateUrl: 'templates/contacts.html',
    controller: 'ContactsCtrl'
  });
  $urlRouterProvider.otherwise('/');
});
```

Let's consider what is going on here, particularly in the `config` function. Here, we set up the core navigation settings for our app by configuring the routing module exposed by `$stateProvider` and `$urlRouterProvider`. By default, Ionic uses the `ui-router` (for more information, visit `https://github.com/angular-ui/ui-router`). This router is state-oriented. That is, it lets you structure your app as a state machine, where each state can be connected to a path and a set of views and controllers. This setup makes it very easy to work with nested views, which is a frequent case when developing for mobile devices, where navigation elements like tabs, side menus, and the like are very common.

Knowing this, let's consider what this code actually does:

1. The `config` function itself takes two arguments, `$stateProvider` and `$urlRouterProvider`. Both are configuration interfaces belonging to `ui-router` and can be used to configure the router when the app bootstraps.

2. We use the `$stateProvider` argument in order to add the state contacts to our app. When the application is in this state, we make the following properties hold:

   ° The current path within the app is root (/),that is, we are in this state whenever we are at the initial path of the app.

   ° The template to be loaded for this state is `templates/contacts.html`, which is the same view template that we created earlier.

   ° The controller associated with this view is `ContactsCtrl`, which was defined earlier as well. Since our `phonebook.controllers` module is loaded as a dependency for our app (see the first line of `app.js`), we only need to name the controller, and the dependency injection system will do the rest.

3. Finally, we configure a default route for our app. This is the route our navigation will resolve to if no other valid route is provided. In our case, we always default to the start screen.

That's all we need to get the routing going! Next, we need to make sure that the `index.html` file has the necessary content to load all the files that we configured so far, including the new contacts view. Open it, and make sure that it has the following content:

```html
<!DOCTYPE html>
<html>
  <head>
    <meta charset="utf-8">
    <meta name="viewport" content="initial-scale=1,
    maximum-scale=1, user-scalable=no, width=device-width">
    <title></title>

    <link href="lib/ionic/css/ionic.css" rel="stylesheet">
    <link href="css/style.css" rel="stylesheet">

    <!-- ionic/angularjsjs -->
    <script src="lib/ionic/js/ionic.bundle.js"></script>

    <!-- cordova script (this will be a 404 during development)
    -->
    <script src="cordova.js"></script>

    <!-- your app's js -->
    <script src="js/controllers.js"></script>
    <script src="js/app.js"></script>
  </head>
  <body ng-app="phonebook">

    <ion-nav-bar class="bar-stable">
      <ion-nav-back-button></ion-nav-back-button>
    </ion-nav-bar>

    <ion-nav-view></ion-nav-view>

  </body>
</html>
```

Pay attention to the highlighted parts:

- The `script` tag simply imports the `phonebook.controllers` module.
- The `ion-nav-bar` tag creates a standard navigation bar, which displays content specific to the current navigation context that the app is in.

If you own an Android device, this will be similar to the action bar at the top, where you have your app logo, the name of the current view, and so on

Likewise, if you own an iPhone, this bar will be the bottom bar, which commonly holds the app's navigation tabs

- The `ion-nav-back-button` tag creates a button to go backwards from the current navigation context to the previous one, much like a back button in a browser.

- Finally, `ion-nav-view` is a special tag, which tells AngularJS where the routing system should bind the templates. In our case, this is where the `templates/contacts.html` template will be rendered when our navigation context is the contacts state, as we defined in our router `config` earlier.

Testing the connection

This is all we need for a basic setup of our app. To make sure it runs, let's try it out in the emulator. From the root of your project folder, run the following in a terminal or command line:

```
ionic serve
```

You should see the following output of the preceding command:

Depending on the browser you run your emulator in, you may not see the top or bottom bar. This is not generated by Ionic, but rather by a Chrome plugin, which automatically sets the size of the window to match that of an iPhone 6. The plugin is called **Mobile/Responsive Web Design Tester**, and we recommend that you try it for your own projects.

Creating the model

We now have a working view and controller. Next, we need a model that can provide the data we need in response to user input.

Services

In AngularJS, it is considered good design practice to keep as little logic in the controllers as possible. Remember that controllers should only be the *glue* between the view and the model. They should ideally not be responsible for doing any of the heavy fetching and crunching of data. For that, AngularJS provides the services.

Services are objects that are injected into other components of your app on demand by the dependency injection system. If you look back at the `app.js` code, you have seen some of them already; both the `$stateProvider` and `$urlRouterProvider` arguments fall in this category.

If you look at advanced AngularJS apps, you will start seeing services pretty much everywhere. They make up a large part of almost any app and can be used to contain almost any kind of functionality. For example, we can create a service that encapsulates access to a given REST API, allowing us to query it by a set of utility functions while the service itself handles connections to the server, security, and so on. Likewise, we can define a service that represents a set of mathematical operations, which can be fed simple data in order to get arithmetic results.

Services are **singletons**, which means that only a single instance of each exists during runtime.

Implementing your business logic in this fashion is important for several reasons, some of which are as follows:

- It provides modular interfaces to work with a certain aspect of your app's business logic. Your model can be built around several services, each providing a single, essential feature of your app's functionality.

- It is efficient, since you only have a single instance of each service available throughout your entire app.

- It makes it easier to extend your app, as units of functionality can be defined and injected wherever they are needed.

Now that we have established how awesome services are, the natural question is, how are we going to go about their creation?

Creating services

AngularJS provides several ways to create services. These ways are referred to as recipes. There are five of them in total—constants, values, providers, factories, and services. Each varies in complexity and the use cases that are suitable for them:

- **Constants**: The most simple service, this is used to define a single constant, which is available throughout the entire application. It varies from the other four in the sense that it is immediately instantiated when an app starts up, which means that it can be used during the configuration phase of the app's lifecycle. Constants are often used to contain constant values such as base URLs.

- **Values**: This is similar to constants, with the notable exception that it is not available during the configuration phase of the app. It may also be used in `decorators`.

- **Factories**: Whereas constants and values are used to store simple values, factories begin making things much more interesting:
  - They provide factory functions, which can be used to define logic rather than just values. This means that a factory can provide multiple functions, which compute values based on input.
  - Factories can have dependencies, which means that you can construct them using other services.
  - Factories are lazily instantiated, which means that they are only instantiated when they are needed.

  For our app here this is the recipe that we will be using, and as such you will see its example soon. In fact, we contend that factories will fit most of the available use cases for most apps.

- **Services**: Services are very, very similar to factories. While there are some minor differences in semantics, their real contribution is to provide a concise syntax. Using a factory or service is almost always a matter of preference. Oh, and in case you are wondering about the name, the AngularJS developers themselves regret calling it *services*, likening it to naming your child *child*.

- **Providers**: Finally, the most advanced recipe is the **provider**, which offers the full range of functionality offered by services (no, not the ones that we just mentioned; we are talking about the actual services—the naming was a pretty bad idea, wasn't it?) In particular, Providers allow you to expose the service to configuration during the config phase of the app before the service is actually used during the run phase. It is worth while (and perhaps surprising) to note that the provider is actually the only recipe for services; the previous four are just syntactic sugar simplifying its use. Because it is so complex, it is an overkill for most cases, which is why there are other options to choose from depending on how complex your model logic is.

So much knowledge, so little space! Let's put what we have learned to good use by actually creating a factory to retrieve contacts.

Creating a factory

Like controllers, it is customary to place your service definitions in their own file (or files, if you prefer to have each service recipe in its own file. Here, we use a common file for all of them). In your project directory, create the following file:

```
js/services.js
```

In this file, put the following:

```
angular.module('phonebook.services', [])
.factory('contactsFactory', function contactsFactory() {
  return {
    all: function () {
      return [];
    }
  }
});
```

Let's have a look at what we have done:

We created a module named phonebook.services to host our services

We defined a basic factory service named contactsFactory

The service which exposes a single utility method is called all, which currently does not do anything (we will change this soon, don't worry).

Now, we need to modify the app.js and index.html files in order to make the app aware of the new service. Make sure that the app.js file starts with the following:

```
angular.module('phonebook', ['ionic',
'phonebook.controllers','phonebook.services'])
```

This injects the services module into the main app. Now, we just need to import the file the module is located in. To do so, make sure that your JavaScript imports in `index.html` look like this:

```
<!-- your app's js -->
<script src="js/controllers.js"></script>
<script src="js/services.js"></script>
<script src="js/app.js"></script>
```

That's it! We now have the groundwork of our app in place. However, if you run the emulator again, you will notice that not much has changed. We are still greeted by the same blank screen. Now, it is time to add some actual content to our app by loading the contact list.

Accessing the device data

Now that the basics of our app are have been implemented, it is time to add some serious data to it. In our case, we want to load the contacts stored on the device that our app is running on so that we can show them in the list that we created earlier.

Accessing native services

You may recall that Ionic is built on top of the Cordova platform, which provides the core interaction with the underlying operating system and hardware. In order to access native services, such as the contact list, we will frequently have to make use of Cordova directly.

In this particular case, we are in a very easy spot; Cordova not only has a full-fledged plugin to interact with the contacts, but also sports a very convenient CLI method to install it.

Go to your project directory and run the following from a terminal or command line:

cordova plugin add org.apache.cordova.contacts

This will install the Cordova Contacts plugin, which will be placed in the following folder:

```
plugins/org.apache.cordova.contacts
```

Feel free to inspect the files before we move on. Next, we need to integrate this plugin with Ionic so that we can use it in our app.

ngCordova

Cordova itself knows nothing about either Ionic or AngularJS. So, accessing its services in an Angular fashion will often require wrapper code. Fortunately, there is already an extensive library for this exact end—**ngCordova**.

To install it, go to the root of your project folder and run the following from a terminal:

```
bower install ngCordova
```

This will install everything we need. Next, let's again import it into our app by modifying the app.js and index.html files. In app.js, make sure that your app dependencies now include ngCordova, as follows:

```
angular.module('phonebook',
  [
    'ionic',
    'ngCordova',
    'phonebook.controllers',
    'phonebook.services
])
```

Likewise, in index.html, make sure that we import the corresponding JS library, as follows:

```
<!-- AngularJS bindings for Cordova -->
<script src="lib/ngCordova/dist/ng-cordova.min.js"></script>
```

Adding Cordova contacts to our factory

The last step here is to integrate the Cordova contacts with the contactsFactory service in order to let it serve the contacts available on the device. Open the js/services.js file and make sure that it contains the following:

```
angular.module('phonebook.services', [])
.factory('contactsFactory', function contactsFactory($q,
$cordovaContacts) {
  /**
  * Turns a raw contact into something more suitable for viewing.
  *
  * @paramrawContact
  */
  function processContact(rawContact) {
    return {
```

```
            name: rawContact.name ? rawContact.name.formatted : '',
            number: rawContact.phoneNumbers ?
            rawContact.phoneNumbers[0].value
        }
    }

    return {
        all: function () {
            // It may take some time to fetch all contacts, so we defer
            it.
            var deferred = $q.defer();
            $cordovaContacts.find({multiple: true}).then(function
            (contacts) {

                varprocessedContacts = [];
                contacts.forEach(function (contact) {
                    processedContacts.push(processContact(contact));
                });

                deferred.resolve(processedContacts);
            }, function (error) {
                deferred.reject(error);
            });
            return deferred.promise;
        }
    }
});
```

The important parts are highlighted. Let's figure out what is going on:

- We inject the following dependencies for our factory:

    - $q: This is the AngularJS service that is used to work with
      promises. This will allow us to create deferred functions,
      which resolve to a value at a later stage. At that point,
      something can be done with their results.

    - $cordovaContacts: This is the Cordova contacts plugin itself
      that is wrapped by ngCordova.

- Since raw contact data is rather clunky, we define a utility method in order
  to process it into something more lightweight.

- We create a promise, which is returned to whoever uses our service. The
  promise in this case is that at some point, we will deliver either a list of
  contacts, or an error indicating why we could not do so.

- We invoke the `find()` function of `cordovaContacts`, which is very similar to the MongoDB function with the same name; it simply returns all the available contacts.

- If we can get the list of contacts, we resolve the `promise` and hand over the list to the caller.

- If we cannot get the list of contacts, we reject the promise, indicating that we were not able to get what was requested. An error message created by `cordovaContacts` is returned to the user.

Our service is now all configured and good to go. There is still one major hurdle though. Emulators have no contacts for us to display! In order to move on, we will first have to take a detour and see how to deploy the app on physical devices before we finally wrap up our app by showing the contacts to the user.

Building for native devices

So far, everything we have done in Ionic has been tested in emulators or remote services. Now that we have an app that uses real phone features, it is time we finally went all the way and built the ultimate end-product—a mobile app.

Building and deploying Ionic apps on physical devices is remarkably easy thanks to the Ionic CLI. We will demonstrate how to do so in the following section.

Android

From the root of your project folder, run the following:

```
ionic platform add android
```

This will add all the files that are necessary to deploy your app on an Android device.

Next, we have two options for the running of our app—an actual Android emulator or a physical device.

Emulator

To run the app in the emulator, first build the project by running the following:

```
ionic build android
```

Finally, start the emulator and deploy the app by running the following:

```
ionic run
```

A physical device

Running an app on a physical device is just as simple as running the app on the emulator. First, connect an Android device that can accept APK installations over USB (see *Chapter 1*, *Setting Up Your Workspace*, for instructions on setting this up). Once this is done, run the following:

```
ionic build android
```

Finally, run the following to install and run the APK on your connected device:

```
ionic run
```

Once the app is deployed, it will start automatically on your device.

The list view revisited

We will now add the finishing touches to our app, as follows:

1. Use logic to display the contacts that we pulled from the contact list
2. Add the pull-to-refresh feature in order to enable users to dynamically refresh the list of users.

First, let's modify the `contacts.html` file in order to handle the rendering of the list itself. Open the file and make sure that it looks like this:

```html
<ion-view view-title="contacts">
  <ion-content class="has-header">
    <ion-refresher
      pulling-text="Pull to refresh"
      on-refresh="doRefresh()">
    </ion-refresher>
    <ion-list>
    <ion-item collection-repeat="contact in contacts"
    type="item-text-wrap">
      <h2>{{contact.name}}</h2>

      <p>{{contact.number}}</p>
    </ion-item>
    </ion-list>
  </ion-content>
</ion-view>
```

Most things look the same, but we have highlighted some important changes:

- We added an `ion-refresher` tag, which creates a pull-to-refresh interface for our view. When the user swipes a finger downward over the screen, the text **Pull to refresh** will be shown. If the full gesture is then carried out (that is, swipe and drop), the `doRefresh()` function, which was defined in the scope of this view, will be called. We will define this function in just a bit.

- We added a `collection-repeat` attribute to the `ion-item` tag. This is a variation of the `ng-repeat` AngularJS attribute, which means that one `ion-item` will be created for each contact number in the contacts collection. The contacts collection needs to be defined in the scope of the view, which will be done next.

Modify the `controllers.js` file to make it look like the following code:

```
angular.module('phonebook.controllers', [])
.controller('ContactsCtrl', function ($scope, contactsFactory) {
  // List of contacts
  $scope.contacts = [];

  $scope.doRefresh = function () {
    contactsFactory.all().then(
      function (contacts) {
        $scope.contacts = contacts;
        $scope.$broadcast('scroll.refreshComplete')
      },
      function (error) {
        alert(error);
        $scope.$broadcast('scroll.refreshComplete')
      },
      function (notify) {
        console.log("Just notifying");
      });
  };
});
```

Let's consider what happened here:

- We created and bound the empty contacts list to the `scope` tag injected into this controller. This corresponds to the same contacts list that the `collection-repeat` directive in our view uses.

- Likewise, we bound the `doRefresh` function, which we already saw in the view, to our scope. We made it do the following:
  - ° Call the `all()` method of the `contactsFactory` class. This gives us a promise that a list of contacts will be delivered at some point in the future.
  - ° If the promise is fulfilled, we bind the resulting list to the scope. Angular will respond to this change by refreshing the view in order to accommodate this change in the model and populating the list with contact information using collection-repeat.
  - ° If the promise fails, we display an error message.

    Angular promises allow us to listen to progress notifications from promises. We do not use this feature here, but we simply catch such messages.

That's it! You should now be able to run your app and browse your contacts. Go through the build steps for native devices again and try it out!

Summary

In this chapter we covered a lot of ground, going into great depth as regards AngularJS and learning more about the interaction of a model, view, and controller. We also saw how to use Cordova plugins and ngCordova in order to access native features (something that we will be doing a lot of in the future chapters). Finally, we saw how to create services and use them in order to serve data to our users.

We also recommend the usage of your favorite browser's inspect element tools, which can give you an insight into any display errors that you might run into when running the application.

9
Accessing Native Phone Features

The main thing that sets hybrid apps apart from ordinary, mobile-friendly web apps is the ability to interact with the operating system and hardware of the underlying mobile device. Modern devices offer a plethora of services to app developers, from GPS and database functionality to Bluetooth, NFC, and other communication technologies. Making good use of these services allows us to build apps that meet the needs of mobile users in the best way possible.

In this chapter, we will continue building on the brief introduction to mobile services that we saw in the last chapter, and we'll do some refreshing as necessary. Our goal is to use the GPS receiver, which is one of the most ubiquitous smartphone features, in order to build a simple navigation app. In doing so, we will also get familiar with a new, fundamental AngularJS component called the directive.

Creating the project

We will start off by setting up the basic structure of our app. As before, we will go for a blank project and build our app from scratch to make sure that we understand how everything works:

1. Create a new project folder for your app. Next, enter into the folder and execute the following from your terminal or command line:

   ```
   ionic start superNav blank
   ```

2. Ionic will now download and configure everything you need in order to deploy a basic app (albeit not a very interesting one). You can see what it looks like by going into your project folder and executing the following command:

```
ionic serve -l
```

The output of this command is shown in the following screenshot:

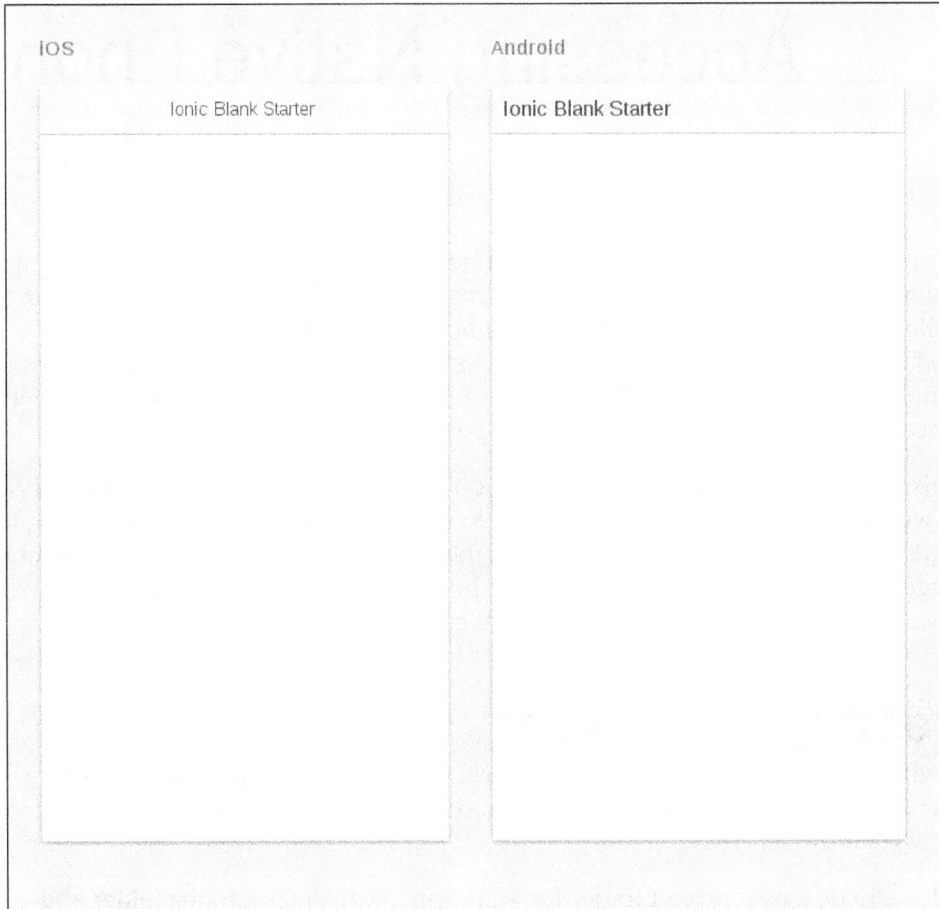

Now that we have the basics in place, let's start adding some basic functionality.

Creating the basic app structure

We want to keep our app as simple as possible—a single screen with a map, together with a toolbar where we can place buttons for various utilities, such as finding the user's current location.

Let's create a basic view that meets this requirement.

Open your app's index.html file and make sure that it looks like the following:

```
<!DOCTYPE html>
<html>
  <head>
    <meta charset="utf-8">
    <meta name="viewport" content="initial-scale=1,
    maximum-scale=1, user-scalable=no, width=device-width">
    <title></title>

    <link href="lib/ionic/css/ionic.css" rel="stylesheet">
    <link href="css/style.css" rel="stylesheet">

    <!-- ionic/angularjs js -->
    <script src="lib/ionic/js/ionic.bundle.js"></script>

    <!-- cordova script (this will be a 404 during
    development) -->
    <script src="cordova.js"></script>

    <script
    src="https://maps.googleapis.com/maps/api/js?key=
    AIzaSyB16sGmIekuGIvYOfNoW9T44377IU2d2Es&sensor=true"></script>

    <!-- your app's js -->
    <script src="js/app.js"></script>
  </head>

  <body ng-app="supernav" ng-controller="MapCtrl">
    <ion-header-bar class="bar-stable">
      <h1 class="title">SuperNav</h1>
    </ion-header-bar>

    <ion-content scroll="false">
```

```
        <div id="map"></div>
    </ion-content>

    <ion-footer-bar class="bar-stable">
    </ion-footer-bar>
  </body>
</html>
```

The browser preview should now look like this (if you closed the server after the previous step, feel free to start it up again and leave it running; it will automatically load any changes made to the underlying sources):

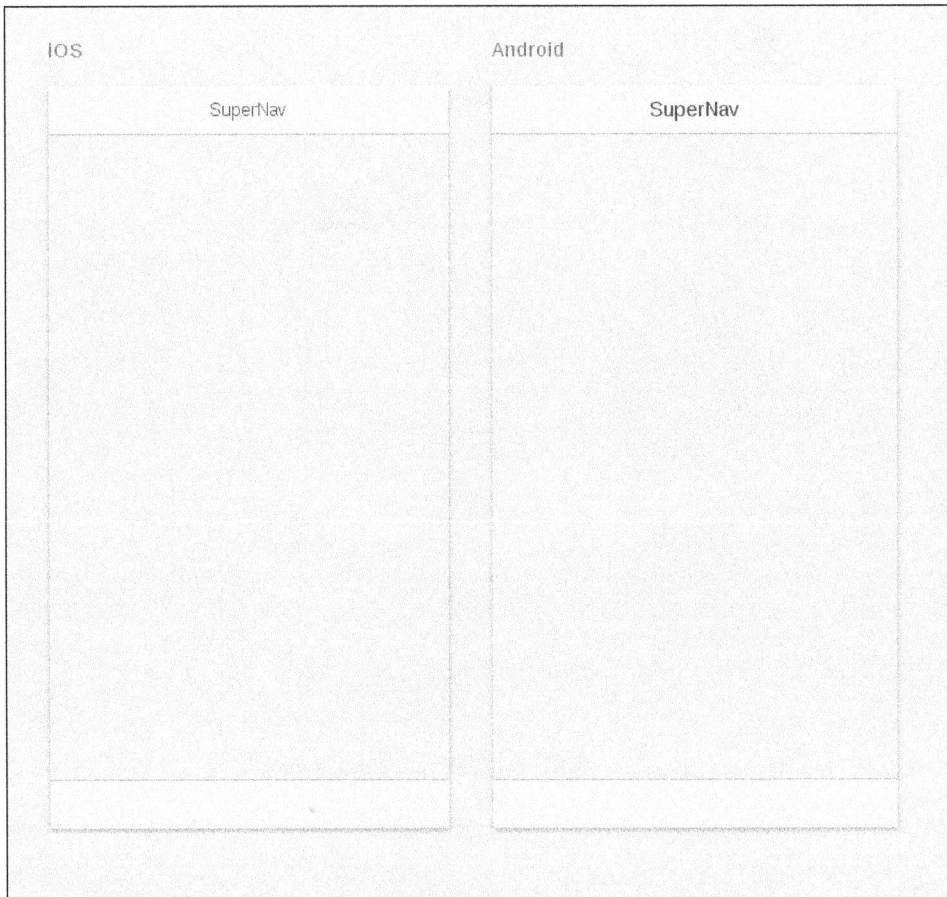

A little bit more content, but nothing exciting as of yet. Have patience; we are getting there.

Integrating Google Maps into the app

Next, we want to integrate the most essential feature of our UI—the map. To do so, we will use **Google Maps**, one of the most popular map services.

If you have ever used the Google Maps application on a mobile device or checked directions to a location on Google, you are already familiar with what Google Maps looks like and some of what it can do. In essence, Google Maps is a complete package that offers everything—scalable maps, satellite imagery, geocoding, and much more. For our purposes here, it is perfect.

To integrate Google Maps into the app , we need to import the Google Maps JavaScript SDK, which is freely available. To do so, add the following import to the `index.html` file:

```
<script
src="https://maps.googleapis.com/maps/api/js?sensor=true">
</script>
```

Next, we will need to designate an area of the UI where the app should be drawn. Change the existing `div id` tag to the following:

```
<div id="map" ng-controller="MapCtrl"></div>
```

In order to render the map properly, we will need to add some custom CSS to force the map to fill its parent container. Open the `www/css/style.css` file and add the following to it:

```
#map {
  display: block;
  width: 100%;
  height: 100%;
}

.scroll {
  height: 100%;
}
```

Also note that we added a binding for a controller for our map. We will use this in order to perform the initial configuration needed in order to render and work with the map. So, let's go ahead and add it! Create the `www/js/controllers.js` file in your project and make sure that it contains the following:

```
angular.module('supernav.controllers', [])
.controller('MapCtrl', function ($scope) {
```

```
    $scope.mapCreated = function (map) {
      $scope.map = map;
    };

    function initialize() {
      var mapOptions = {
        center: new google.maps.LatLng(57.661577, 11.914768),
        zoom: 16,
        mapTypeId: google.maps.MapTypeId.TERRAIN
      };

      $scope.map = new google.maps.Map(
        document.getElementById('map'), mapOptions
      );

      $scope.onCreate({map: map});

    }

    if (document.readyState === "complete") {
      initialize();
    } else {
      google.maps.event.addDomListener(window, 'load', initialize);
    }
  });
```

Here, we defined a new `supernav.controllers` module, which will contain the controllers of our app. For now, it only has one such controller — `MapCtrl`. Let's go through it and consider what it does:

1. We first defined the map scope variable, which will be used to simply refer to the map that we are working with. We also defined a scope function in order to bind a value to this variable.

2. We defined the `initialize` local function, which will be used in order to set up and configure a Google Maps instance as follows:

   ○ Here, we defined the `mapOptions` object, which provides the initial settings for the map to be created. It has the following properties:

center: This property includes latitude and longitude coordinates for the point on the Earth's surface on which the map will initially be centered on. The coordinate pair is passed as an instance of Google Maps' own LatLng object.

zoom: This is the degree of zooming the location that needs to be applied to the map.

mapTypeId: This is the kind of map that we want to show. Google Maps supports several different view modes, ranging from raw satellite imagery to detailed views of roads, businesses, and more.

° We then created the actual Map object. We pass the following to its constructor:

This is the part of the **Document Object Model (DOM)** where we want to bind the map. In our case, it is the div with the map ID, which was earlier defined in our view. We used the standard getElementById DOM function in order to retrieve a reference.

Then we have mapOptions that we defined earlier.

Finally, after the map has been created, we bind it to the map scope object that we defined earlier.

3. Finally, if the DOM is fully loaded, we attempt to execute the initialize function that we just defined. If the DOM is not ready yet, we instead register it as a callback that needs to be run once it is.

4. All we need to do now is make sure that the controller is properly loaded and put in charge of the map. To do so, first make sure that the JavaScript file is imported by adding the following to your index.html file:

```
<script src="js/controllers.js"></script>
```

5. Next, modify the app.js file in order to make sure that the module is listed as a dependency, as follows:

```
angular.module('supernav', ['ionic',
'supernav.controllers'])
```

That's it! The browser preview should now look like this:

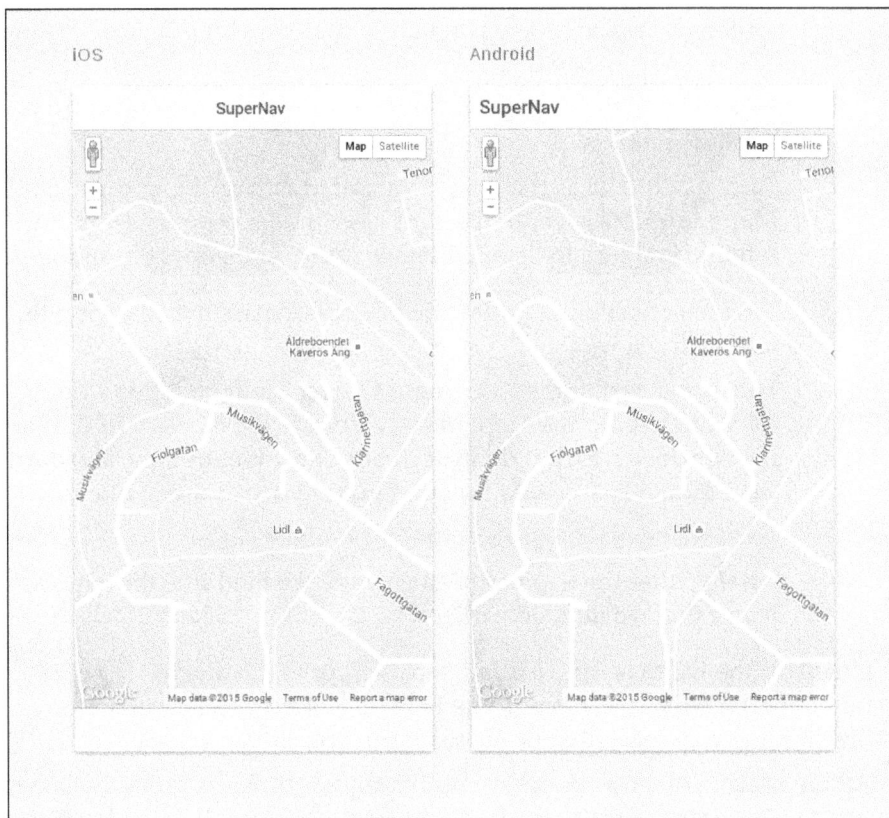

We have come pretty far already. While we do not have any advanced navigation capabilities yet, we have successfully built a basic app that people can use just for the purpose of browsing the maps of the world. Not bad for work that took just 10 minutes!

Before we move on though, it is worth pausing and considering the architectural road that we have travelled so far. Everything we have done here is standard AngularJS practice — create a `view` for the element that we want to display (in this case, a `map`), create a `controller` for it and some logic, and activate the controller by integrating the map into the app. However, you may recall that we mentioned earlier in the book that the the greatest advantage of AngularJS is the creation of enhanced HTML. We have already seen how this works through data binding, live DOM updates, and other things. However, AngularJS also offers us the ability to define custom HTML tags in order to define elements, which can be reused in several parts of the application. Our map, which we created here, is a good candidate. What if we could just encapsulate it in a `<map>` tag? We can, and to get there, we need to talk about directives.

Angular directives

Simply put, directives are custom HTML elements. You write them like ordinary HTML elements, but their functionality is defined entirely programmatically. Thus, they extend the standard HTML syntax by letting us add whatever we need to it in order to build truly dynamic pages.

Creating directives

Like the services and controllers that we have already seen, directives are defined as components of modules, and AngularJS gives us the tools that are necessary to create them. We will explore this process by creating a directive for the map that we created earlier.

The first thing that we will need to do is create a new file named www/js/ directives.js for the directives of our project. Create this file and add the following to it:

```
angular.module('supernav.directives', [])
.directive('map', function () {
  return {}
});
```

The directive module function is used to define a directive for a module, and as you might have guessed, its first parameter is the name of the directive itself, while the second one is a factory function that gives us an object describing how the directive works. In that sense, directives are similar to the services that we studied earlier.

Restricting directives

Let's start building the factory function for our map directive. The first thing that we should do is add a restriction to the directive in order to tell the AngularJS parser which kinds of HTML elements this particular directive may occur as:

```
angular.module('supernav.directives', [])
.directive('map', function () {
  return {
    restrict: 'E'
  }
});
```

Right now, you are probably exclaiming, *E? What is this E of which you speak?* Well, AngularJS allows us to confine a directive to the following three different classes of elements:

- **E (Elements)**: These are your standard HTML tags, such as `<map></map>`
- **A (Attributes)**: These are the element attributes, such as `<div map></div>`
- **C (Classes)**: These are the customized element class attributes that are mapped to the directive, such as `<div class="map"></div>`

You are not required to stick with just one restriction. For example, you can also write the following in order to restrict it to elements and attributes:

```
angular.module('supernav.directives', [])
.directive('map', function () {
  return {
    restrict: 'EA'
  }
});
```

Hence, the Angular parser will detect the directive if you write either `<map></map>` or `<div map></map>`.

> You will frequently find that it makes sense to restrict directives only to a single kind of element. This is good practice as it reduces the complexity of your app.

Scope isolation

Just like controllers, directives are able to access the scope in which they are operating. However, it is also possible (and generally considered good practice) to create an isolated scope for the directive. This scope will contain a set of data that only the current instance of the directive is aware of. In addition to this, scope isolation also helps you create reusable widgets, which enhance code quality.

We achieve this by defining scope injection points in our `directive`, which will take the form of the standard HTML attributes:

```
angular.module('supernav.directives', [])
.directive('map', function () {
  return {
    restrict: 'E',
```

```
      scope: {
        onCreate: '&'
      }
  });
```

Here, we defined an injection point called `onCreate`, which maps the directives to a `function` in the parent scope that we are isolating (the `&` symbol signifies a binding by delegation). For example, let's say that we want to inject the `onCreate` method from `MapCtrl` into the isolated scope. We will then write our directive like this:

```
<map on-create="mapCreated(map)"></map>
```

At this point, the `map` parameter is not bound. Later, we will see how to define and pass it to the function from within the `directive` itself in the next section.

However, before we move on, did you observe that although we name our injection point `onCreate`, we wrote it as `on-create` in the actual HTML? This is due to an AngularJS process called **normalization**. Through this, attributes and tags are translated into a more concise form. Part of the process involves replacing **hyphen-bound** words with **camel-cased** words. We will give you the reference to the documentation if you wish to know more about how it works, since understanding it is not crucial to developing our directive here.

DOM manipulation

Ultimately, we want our map directive to expand and show a map where it occurs in the DOM. To do so, we will need to allow it to actually manipulate the DOM itself.

The typical way to achieve this is by providing the `directive` with a `link` function, which allows it to look into the DOM update process. Let's add one link to our `map` directive, as follows:

```
angular.module('supernav.directives', [])
.directive('map', function () {
  return {
    restrict: 'E',
    scope: {
      onCreate: '&'
    },
    link: function ($scope, $element, $attr) {
      function initialize() {
        var mapOptions = {
```

```
        center: new google.maps.LatLng(43.07493, -89.381388),
        zoom: 16,
        mapTypeId: google.maps.MapTypeId.ROADMAP
      };
      var map = new google.maps.Map($element[0], mapOptions);

      $scope.onCreate({map: map});

      google.maps.event.addDomListener(
        $element[0], 'mousedown', function (e) {
          e.preventDefault();
          return false;
      });
    }

    if (document.readyState === "complete") {
      initialize();
    } else {
      google.maps.event.addDomListener(window, 'load',
      initialize);
    }
    }
  }
});
```

Looks oddly familiar, doesn't it? This is the same `initialize` function and associated map setup procedure that we defined in our `controller` earlier, albeit with some slight modifications. We have already covered how this works. So, let's go over how it figures in the context of the `link` function:

The `link` function takes the following three parameters:

- `$scope`: This is the scope under which the directive is rendered.

- `$element`: This is the tag to which the directive is bound, which is `<map>` in our case. The tag is wrapped in the JQuery-like jqLite library, which allows us to perform direct manipulations on it.

- `$attr`: This defines the attributes for the directive element along with their associated values.

Inside the `initialize` function itself, we now use `$element[0]` in order to get the name of the tag itself (map in our case). We also use the `$scope` parameter in order to call the `onCreate` delegate in the parent scope (note how we explicitly need to define the parameter name and its associated value in this case).

Putting it all together

We now have a full-fledged directive, and it's time to integrate it into our app. First, make sure that the new directive is properly loaded. The first line of the app.js file should look like this:

```
angular.module('supernav', ['ionic', 'supernav.controllers',
'supernav.directives'])
```

Likewise, the index.html file should contain the following import:

```
<script src="js/directives.js"></script>
```

Next, make sure that the ion-content section in index.html now looks like this:

```
<ion-content scroll="false">
  <map on-create="mapCreated(map)"></map>
</ion-content>
```

Next, since we moved the rendering logic for the map into the directive, remove it from the controller.js file, which should now look like this:

```
angular.module('supernav.controllers', [])
.controller('MapCtrl', function ($scope) {
  $scope.mapCreated = function (map) {
    $scope.map = map;
  };
});
```

Finally, we need to make some slight modifications to style.css in order to make sure that the map directive will render on app properly. Make sure that it looks like this:

```
map {
  display: block;
  width: 100%;
  height: 100%;
}

.scroll {
  height: 100%;
}
```

That's it! Ensure that you reload the preview in your browser if necessary. It should look just the same as it did when we were not using a directive. We have succeeded in putting all together!

Adding geolocation via GPS

Now that we have an even better working map view, let's go ahead and add some basic navigation features to it, namely, the ability to focus the map on our current physical location.

First, let's add a `button` to the footer toolbar for now. Make sure that the `ion-footer-bar` tag looks like the following:

```
<ion-footer-bar class="bar-stable">
  <a ng-click="centerOnUser()" class="button button-icon icon
ion-navigate"></a>
</ion-footer-bar>
```

Your preview should now look like this:

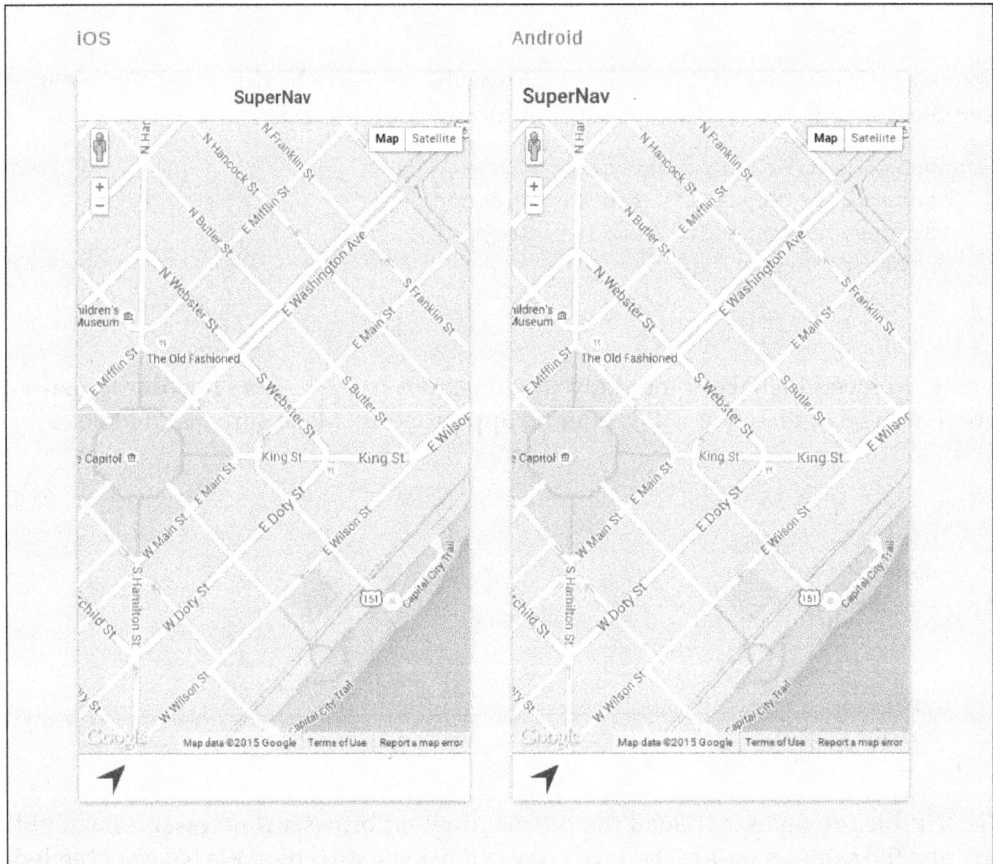

Now, we need to hook into the location capabilities of the native device in order to find the user's current location. Fortunately, this can be done directly through the HTML5 geolocation interface. To see it in action, let's add the following to the `controller.js` file inside the `MapCtrl` controller:

```
$scope.centerOnUser = function () {
  console.log("Centering on user");
  if (!$scope.map) {
    return;
  }

  navigator.geolocation.getCurrentPosition(function (pos) {
    console.log('Got pos', pos);
    $scope.map.setCenter(
      new google.maps.LatLng(pos.coords.latitude,
      pos.coords.longitude));
  }, function (error) {
    alert('Unable to get location: ' + error.message);
  });
};
```

Note what we did here.

We defined the `centerOnUser()` scoped function. This function is in turn bound to the location button that we just defined in the `index.html` file.

In this function, we have the `navigator.geolocation.getCurrentPosition` function, a part of the HTML5 standard, in order to retrieve the current location of the user. This function takes a callback as an argument, which in turn takes a position object, `pos`, as an argument.

Inside the callback, we use the latitude and longitude value stored in `pos` in order to recenter the map using the `setCenter()` method provided by Google Maps.

You can now try out the navigation for yourself in the preview. Clicking on the location button should change the focus of the map to your current location. That's it. We are done!

Summary

In this chapter, we worked directly with the Google Maps API in order to render the map and work with maps. While this is perfectly fine, we also recommend that you check out the module that can be found at `http://angular-ui.github.io/angular-google-maps/#!/`, which makes it even easier to work with Google Maps in AngularJS.

Apart from the navigation features that we covered here, there are of course a plethora of services offered by mobile devices. Fortunately, almost all of the services have a corresponding Cordova plugin, which will let you make use of these services. We recommend that you refer to the Cordova plugin registry, which can be found at `http://plugins.cordova.io/#/`, whenever you need to find one.

You can do much, much more with Google Maps than what we explored here. We highly recommend that you explore the developer's page for Google Map, which can be found at `https://developers.google.com/maps/`, to get a better view of the available possibilities regarding Google Map.

The navigation app that we built here is really the `map` starter app in Ionic. You can find the complete source code for it at `https://github.com/driftyco/ionic-starter-maps`.

In this chapter, you learned how to use native phone features and Google Maps in order to build a simple navigation app. Most importantly, you also studied directives in AngularJS, a powerful feature that lets you create custom HTML elements.

10
Working with APIs

During the creation process of an application, in today's dynamic living and working environments, connecting your mobile application to a backend will facilitate the process of data storage and connection of data to companion web applications, which will provide users with enhanced capabilities and a satisfactory experience.

Introducing a backend into your project can be facilitated through the use of the `Parse.com` REST API. Parse offers an easy-to-use, straightforward, and scalable platform, which can be implemented within an application based on the Ionic framework.

As we go through this chapter, we will start off by setting up a Parse backend and an Ionic project. After this, we will create a connection between them to achieve our desired effect.

Setting up Parse

The first step that is required to get this project up and running is to visit `www.parse.com` and create an account or log in to that site if you already have an account. In the free plan, you will get a file storage of 20 GB, a database storage of 20 GB, and the ability to transfer up to 2 TB of data.

If you're using `Parse.com` for the first time, you will be welcomed by the following screen:

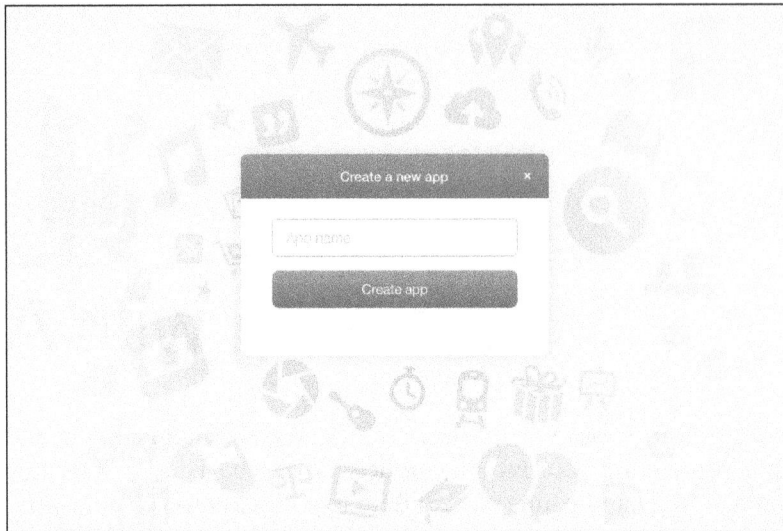

Let's name the app Ionic-ToDo. Once the app has been created, you will see the following screen if this is the first application that you created with Parse. Make sure that you note down the **Application ID** and **REST API Key**:

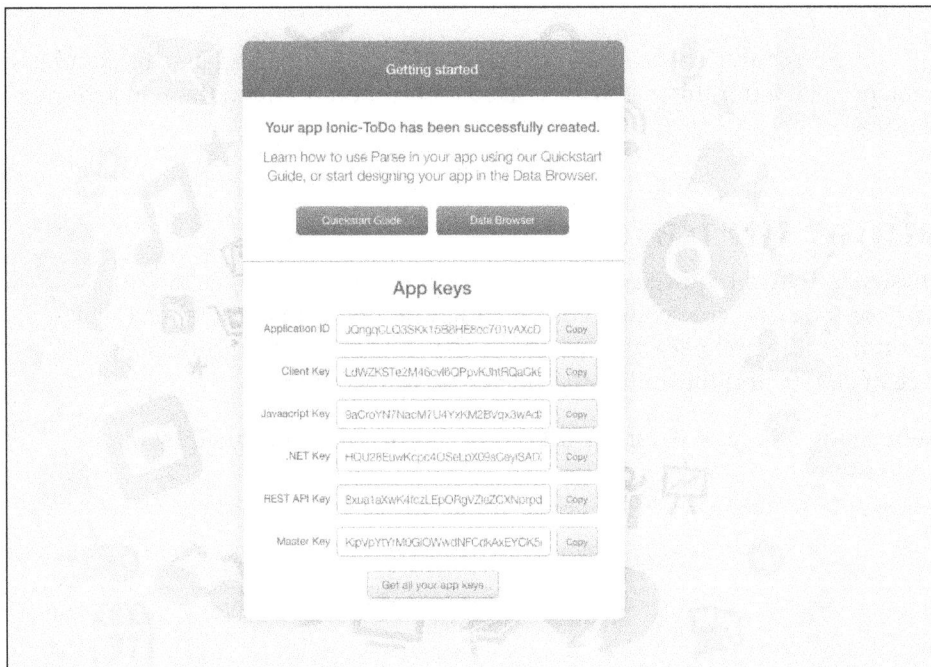

The next part involves proceeding to the **Data Browser**. In this view, click on **Add Class** in the sidebar. The role of a class is to store data in the Parse application instance. In our case, we will name the class Todo and choose **Custom** as the class type:

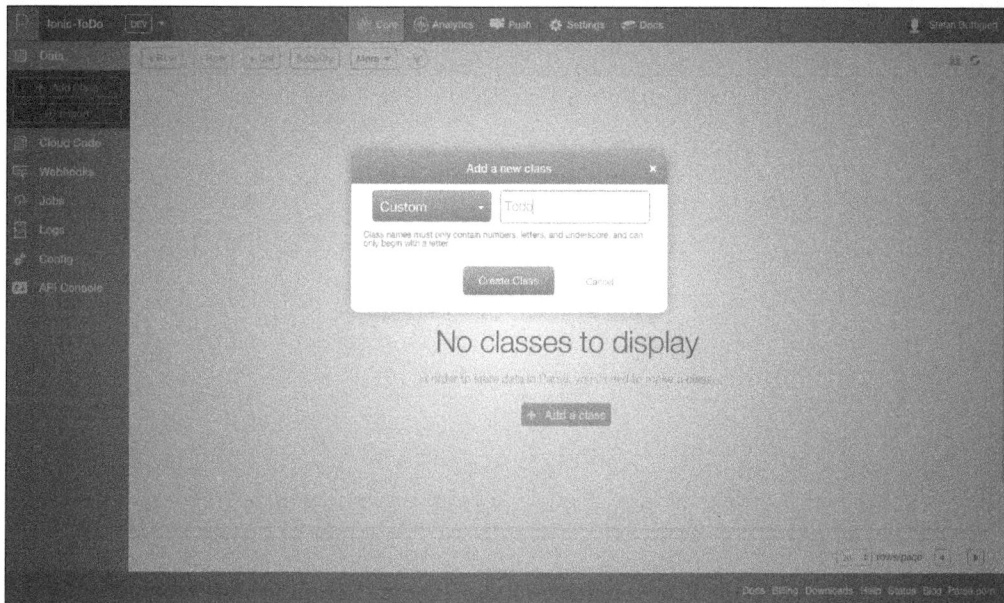

To finalize the class creation process, click on **Create Class**, and you will have your class ready. In the data browser, you'll see that your newly created class already has some properties:

- objectId: This is a unique ID that represents an individual Todo item in the collection
- createdAt: This tells us when the Todo item was added to Parse
- updatedAt: This tells us when the Todo item was last updated

In order to personalize the Todo application and allow for the creation of to-dos, we need a custom property to hold the contents of a Todo item. In the Data Browser, click on **+Col**, and the following dialog box will pop up:

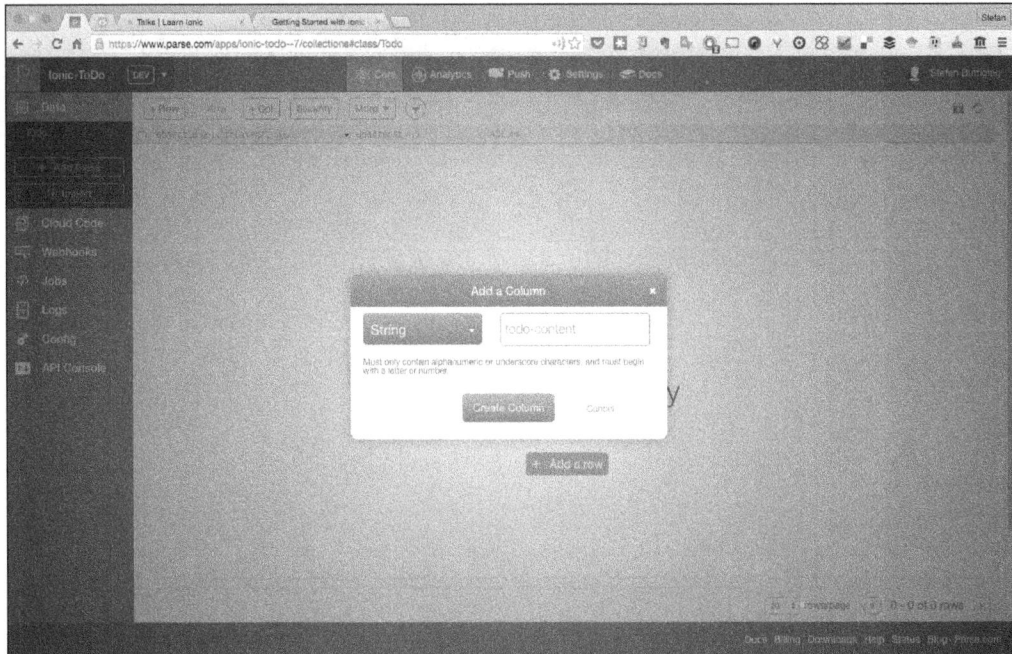

For this property, we will choose **String** as the type of column and input `todo-content` in the name field. Conclude this process by clicking on **Create Column**, and we are done!

Once we have the class ready, we can start creating our Ionic app.

Setting up the Ionic project

We will create the Ionic app by running the following command:

```
ionic start ionictodo blank
```

This will create a blank Ionic starter app named `whichapp`, which will give us the necessary setup that is required to connect the Parse service to our project.

Connecting Parse to our project

In order to connect the Parse data to our project, we will need to create an **AngularJS** service that performs the necessary CRUD operations to interact with the Parse REST API.

The code editor of our choice is Atom, the GitHub open source editor. Start by adding the project folder to Atom by navigating to it:

File | Add Project Folder

Create a file named `services.js` with the following path:

www | js | services.js

Start by connecting the `services`. Define the `service` as follows:

```
angular.module('ionictodo.services',[]).factory('Todo',['$http',
function($http){
  return {


  }
}]);
```

At this stage, our factory object `Todo` is an empty object, and we will need to add the necessary `Parse.com` API methods to it. It's important to note that the hostname is `https://api.parse.com` in all cases. `/1/` means that we are using version 1 of the API.

The following is the factory object with the five required methods:

```
angular.module('ionictodo.services',[]).factory('Todo',
['$http','PARSE_CREDENTIALS',function($http,PARSE_CREDENTIALS){
  return {
    getAll:function(){
      return $http.get('https://api.parse.com/1/classes/Todo'',{
        headers:{
          'X-Parse-Application-Id': PARSE_CREDENTIALS.APP_ID,
          'X-Parse-REST-API-Key':PARSE_CREDENTIALS.REST_API_KEY,
        }
      });
    },
    get:function(id){
    return $http.get('https://api.parse.com/1/classes/Todo/'+id,{
      headers:{
        'X-Parse-Application-Id': PARSE_CREDENTIALS.APP_ID,
        'X-Parse-REST-API-Key':PARSE_CREDENTIALS.REST_API_KEY,
      }
    });
    },
    create:function(data){
```

```
      return $http.post
    ('https://api.parse.com/1/classes/Todo',data,{
      headers:{
        'X-Parse-Application-Id': PARSE_CREDENTIALS.APP_ID,
        'X-Parse-REST-API-Key':PARSE_CREDENTIALS.REST_API_KEY,
        'Content-Type':'application/json'
      }
    });
    },
    edit:function(id,data){
      return $http.put
    ('https://api.parse.com/1/classes/Todo/'+id,data,{
      headers:{
        'X-Parse-Application-Id': PARSE_CREDENTIALS.APP_ID,
        'X-Parse-REST-API-Key':PARSE_CREDENTIALS.REST_API_KEY,
        'Content-Type':'application/json'
      }
    });
    },
    delete:function(id){
      return $http.delete
    ('https://api.parse.com/1/classes/Todo/'+id,{
      headers:{
        'X-Parse-Application-Id': PARSE_CREDENTIALS.APP_ID,
        'X-Parse-REST-API-Key':PARSE_CREDENTIALS.REST_API_KEY,
        'Content-Type':'application/json'
      }
    });
    }
  }
}]);
```

We will also need to declare the PARSE_CREDENTIALS value service, as follows:

```
.value(PARSE_CREDENTIALS 'PARSE_CREDENTIALS',{
  APP_ID: 'yourappid',
  REST_API_KEY:'yourrestapikey'
});
```

Replace yourappid and yourrestapikey with your previously noted application ID and REST API key respectively.

Defining app states

In order to connect the different apps' functions, we will need to define some states for our app, which are as follows:

- `todos`: This lists all the to-do items
- `createTodo`: This allows users to create a new to-do item
- `editTodo`: This allows users to update a to-do item

The app states are defined in the `app.js` file named `whichfile`, which is available at the following path:

www | js | app.js

Replace the existing `app.js` code with the following code:

```
angular.module('ionictodo', ['ionic','ionictodo.
controllers','ionictodo.services'])

.run(function($ionicPlatform,$state) {
  $ionicPlatform.ready(function() {
    // Hide the accessory bar by default (remove this to show
    the accessory bar above the keyboard
    // for form inputs)
    if(window.cordova && window.cordova.plugins.Keyboard) {
      cordova.plugins.Keyboard.hideKeyboardAccessoryBar(true);
    }
    if(window.StatusBar) {
      StatusBar.styleDefault();
    }
    $state.go('todos');
  });
}).config(function($stateProvider){
  $stateProvider.state('todos',{
    url:'/todos',
    controller:'TodoListController',
    templateUrl:'views/todos.html'
  }).state('createTodo',{
    url:'/todo/new',
    controller:'TodoCreationController',
    templateUrl:'views/create-todo.html'
  }).state('editTodo',{
    url:'/todo/edit/:id/:content',
```

```
        controller:'TodoEditController',
        templateUrl:'views/edit-todo.html'
    });
});
```

Creating controllers and templates

Once you have defined the states, you need to create the controllers and provide a template for each of them. Since we would like to edit and create tasks in a list view, we will define the following three controllers:

- `TodoListController`
- `TodoCreationController`
- `TodoEditController`

In order to implement the controllers, we will need to create a new file entitled `controllers.js`, which should be available at the following path:

www | js | controllers.js

In this file, we will declare all our controllers whichcontrollers, which will be done with the help of the following code:

```
angular.module('ionictodo.controllers',[]).controller('TodoListControl
ler',['$scope','Todo',function($scope,Todo){

  Todo.getAll().success(function(data){
    $scope.items=data.results;
  });

  $scope.onItemDelete=function(item){
    Todo.delete(item.objectId);
    $scope.items.splice($scope.items.indexOf(item),1);
  }

}]).controller('TodoCreationController',['$scope','Todo','$state',func
tion($scope,Todo,$state){

  $scope.todo={};

  $scope.create=function(){
    Todo.create({content:$scope.todo.content}).success
    (function(data){
```

```
      $state.go('todos');
    });
  }

}]).controller('TodoEditController',['$scope','Todo','$state','$stateP
arams',function($scope,Todo,$state,$stateParams){

  $scope.todo={id:$stateParams.id,content:$stateParams.content};
  $scope.edit=function(){
    Todo.edit($scope.todo.id,
    {content:$scope.todo.content}).success(function(data){
      $state.go('todos');
    });
  }
}]);
```

For each controller, we require templates to present the controllers. We will start off with the `todo` list controller, which uses `ion-list` to display all the items in the `todo` list. In order to start off with this process, we will create a new folder in www, which will be named views. Within the views folder, we then need to design three new HTML files named create-`todo.html`, edit-`todo.html`, and `todos.html`.

For `todos.html`, we need to declare the `ion-list` UI element as follows:

```
<ion-header-bar class="bar-positive">
  <div class="buttons">
    <button class="button button-icon icon ion-ios7-minus-outline"
    ng-click="data.showDelete = !data.showDelete;"></button>
    <h1 class="title">All Todo Items</h1>
    <button class="button" ui-sref="createTodo">New</button>
  </div>
</ion-header-bar>
<ion-content>
  <ion-list show-delete="data.showDelete">
    <ion-item ng-repeat="item in items | orderBy: '-createdAt'"
    item="item"
      href="#/todo/edit/{{item.objectId}}/{{item.content}}">
      {{item.content}}
      <ion-delete-button class="ion-minus-circled"
      ng-click="onItemDelete(item)"></ion-delete-button>
    </ion-item>
  </ion-list>
</ion-content>
```

The creation of a new `todo` list is presented with a text area and an **Add** button. When you click on this button, the `$scope.create()` method gets called. Navigate to **create | todo.html**. This should be presented in the code as follows:

```
<ion-header-bar class="bar-positive">
  <div class="buttons">
    <button class="button" ui-sref="todos">Back</button>
    <h1 class="title">Create Todo</h1>
  </div>
</ion-header-bar>

<ion-content>
  <div>
    <ion-list>
      <li class="item item-input item-stacked-label">
        <textarea type="text" placeholder="Start typing..."
        name="content" rows="10" ng-model="todo.content"
        required></textarea>
      </li>
      <ion-button class="button button-block button-positive"
      ng-click="create()">
        Add
      </ion-button>
    </ion-list>
  </div>
</ion-content>
```

The ability of editing a `todo` list is a necessary feature in any to-do list app, and in our project, we will create an `edit-todo.html` file that allows the modification of an existing to-do item, as follows:

```
<ion-header-bar class="bar-positive">
  <div class="buttons">
    <button class="button" ui-sref="todos">Back</button>
    <h1 class="title">Edit Todo</h1>
  </div>
</ion-header-bar>

<ion-content>
  <div>
    <ion-list>
      <li class="item item-input item-stacked-label">
        <textarea type="text" name="content" rows="10"
        ng-model="todo.content" required></textarea>
      </li>
```

```
        <ion-button class="button button-block button-positive"
        ng-click="edit()">
          Update
        </ion-button>
      </ion-list>
    </div>
  </ion-content>
```

Testing our application

At this stage, we have produced the basic functionality of our application and we are ready to test the application in the browser.

On your terminal, navigate to the project directory. Once you're in the project folder, run the following command:

```
ionic serve --lab
```

When you execute the command, your browser will automatically load live screenshots of the application running in Android and iOS side by side. The screen will look like this:

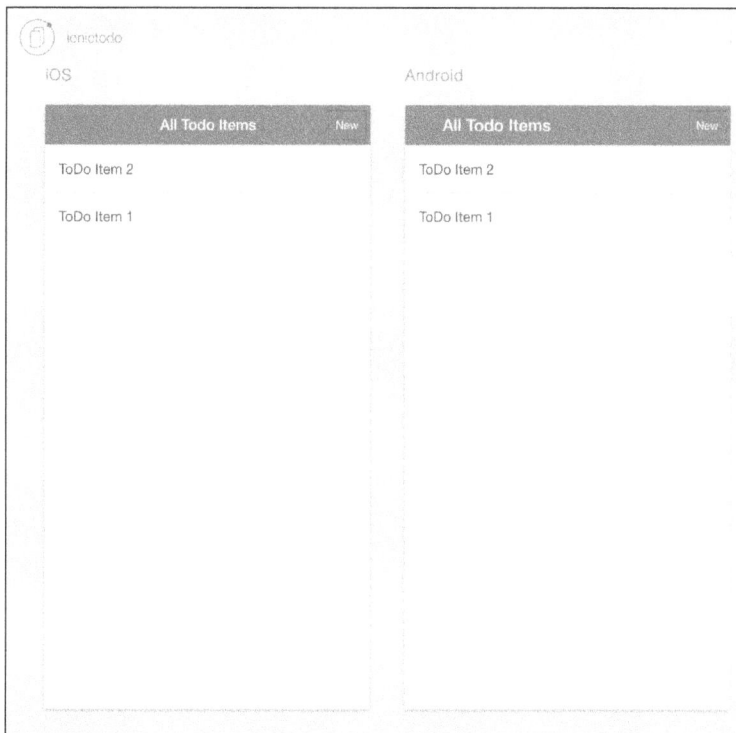

At this stage, once you add new items, they might not show up instantly within your application and you will be required to refresh the page.

Summary

When connecting the app to APIs, the topics that we covered in this chapter are just the tip of the iceberg as regards the possibilities. There are endless opportunities available if you wish to extend such an application, from the perspective of the UI to the API extensibility.

First, we can provide an option of deleting or archiving the completed to-do lists. We can also add an option of a pull-to-refresh feature in order to update lists of to-do items. From the point of view of user experience, we can add more interactive controls, such as swiping to the left or right to mark a to-do item as a completed task. The basis of the knowledge that we have built in this chapter will make you familiar with the necessary concepts that are required to take on more complex tasks, which will be faced by us in the next chapter.

11
Working with Security

User management and conditional access in various forms has become almost ubiquitous in modern apps, and modern users expect the possibility of logging in with an account that they already have or registering and using a new one. Further, they expect that the data that they make available under their accounts will remain secure and in compliance with the applicable privacy legislation.

In this chapter, we will explore how we can add security to our Ionic application, which meets most of these demands. We will start by explaining how routing between views in Ionic works and how we can secure access to individual routes. Finally, we will give some pointers to further reading, including advanced authentication concepts such as **OAuth**.

An overview of client-side security

We will start off by briefly discussing some core concepts that are involved in securing client applications, such as the apps that we are developing in this book.

Client-side security is a convenience

The first thing that you need to know about client-side security is that there really is no such thing as client-side security. The app that you deliver to your users, whether native or hybrid, is exposed to the possibility of tampering, reverse engineering, cracking, and a number of other things that fundamentally compromise its integrity. Thus, you can never really trust client applications with the important part of security in your app, which is ultimately required to safeguard your users' private information and make sure that unauthorized users cannot access data that they should not be allowed to access. In particular, this data is private and cannot be accessed by other users. In fact, many of the most severe blows in terms of security failures of larger companies has been due to user data being compromised and leaked en-masse to unauthorized parties.

However, you can provide security that is good enough for the app's intended use. For example, even if it will not guarantee complete security, you can still attempt to deter less severe privacy invaders from trying to glean personal data from an app by using more advanced security measures such as fingerprint scanners on some devices, or by using encryption on local data while forcing users to pick very strong encryption keys.

The fact that client-side security is not a final measure in safeguarding your users' data should of course not deter you from using it. In fact, client-side security brings a lot of benefits in terms of how we structure our apps. Importantly, it allows us to to create rich user experiences, where the sections of the app accessible to the user can be limited based on the users' authentication status.

The basic components of client-side security

While details may vary across different systems and implementations, there are some fundamental concepts of client-side security that are in use almost everywhere. The following are some of the concepts:

- **Authentication tokens**: These are data that uniquely identify an authenticated user in a system. They are granted by the system itself—or an associated, trusted system—in response to the user providing legitimate authentication information to the system. For example, this information can be a username-password pair, a fingerprint/iris scan, or some other trusted means of authentication.

- **Secure local storage**: In order to improve user experience, we most certainly do not want to force our users to authenticate themselves every single time they use the app. Just imagine a situation where you had to log in again to every single account on your device every time you restarted it! In order to work around this, we use some kind of secure storage, where the access tokens stored under the previous step are preserved. The app itself then simply extracts the token from this storage and uses it in order to perform authenticated communication with the server. The term secure local storage implies some necessary security measures as regards how access tokens are stored and retrieved. This is necessary since a compromised access token will allow an unauthorized party to be masked as the legitimate user. On most mobile devices, there are native features for storage where security is handled by the resident operating system. In other cases, developers can opt to use other solutions, such as an encrypted file storage that require some external mechanism to unlock the system.

- **Secure communication**: Access tokens can be compromised in storage. Furthermore, they are also open to theft while in transit. For example, various types of man-in-the-middle attack, where an attacker is masked as a legitimate endpoint for a network connection, can be used in order to intercept an access token during a transfer in order to steal it and consequently, the user's access privileges. In most cases, secure communication is nothing that you as a developer have to worry about implementing manually. Encrypted connections via HTTPS are increasingly becoming the standard way of communication across the Internet, and they provide very robust security for data over network endpoints. Meanwhile, support for it is present in the network stacks of virtually all major operating systems, both for stationary and mobile devices.

Building a secure app

Now that we have a better understanding of client-side security and its drawbacks, let's put it into practice by developing an app with the following features:

- There is a public home screen that can be seen by everybody who uses the app
- There is a private part that shows some personal information about a user, which is only accessible to authenticated users
- There is logic for the authentication of users through a simple log-in form
- There is logic for the authorization and authentication of users to access the otherwise private parts of the application

Starting off

Let's start with the configuration of our basic project structure. If you have read the book until this point, this should be second nature to you by now! Go to a desired project directory, and from there, just run the following from your terminal or command line:

```
ionic start secureApp
```

This will create a basic, blank Ionic app. Let's add some basic structure to it. The first thing that we want to do is add two basic navigation states—home and public. Navigate to your app's www/js folder and make sure that app.js has the following:

```
angular.module('secureApp', [])
.run(function ($ionicPlatform) {
```

```
    $ionicPlatform.ready(function () {
      // Hide the accessory bar by default (remove this to show
      // the accessory bar above the keyboard for form inputs)
      if (window.cordova && window.cordova.plugins.Keyboard) {
        cordova.plugins.Keyboard.hideKeyboardAccessoryBar(true);
      }
      if (window.StatusBar) {
        // org.apache.cordova.statusbar required
        StatusBar.styleDefault();
      }
    });
  })
  .config(function ($stateProvider, $urlRouterProvider) {
    $stateProvider
      .state('app', {
        url: "/app",
        abstract: true,
        templateUrl: "templates/menu.html"
      })
      .state('app.home', {
        url: "/home",
        views: {
          'menuContent': {
            templateUrl: "templates/home.html"
          }
        }
      })
      .state('app.private', {
        url: "/private",
        views: {
          'menuContent': {
            templateUrl: "templates/private.html"
          }
        }
      });
      // if none of the above states are matched, use this as the
      fallback
```

```
      $urlRouterProvider.otherwise('/app/home');
   });
```

This will set up the essential navigation states for the app, which fortunately are very few at this point! However, we still need to add the necessary templates. Inside the www directory, create a `templates` directory and add the following three files to the path www/templates/menu.html:

```
<ion-side-menus enable-menu-with-back-views="false">
  <ion-side-menu-content>
    <ion-nav-bar class="bar-stable">
      <ion-nav-back-button></ion-nav-back-button>
        <ion-nav-buttons side="left">
          <button class="button button-icon button-clear
          ion-navicon"
          menu-toggle="left">
          </button>
        </ion-nav-buttons>
    </ion-nav-bar>
    <ion-nav-view name="menuContent"></ion-nav-view>
    </ion-side-menu-content>
    <ion-side-menu side="left">
      <ion-header-bar class="bar-stable">
        <h1 class="title">Left</h1>
      </ion-header-bar>
      <ion-content>
        <ion-list>
          <ion-item menu-close
          href="#/app/home">
            Home
          </ion-item>
          <ion-item menu-close
          href="#/app/private">
            Private
          </ion-item>
        </ion-list>
      </ion-content>
    </ion-side-menu>
  </ion-side-menus>
```

The following code snippet represents the `home.html` templates at the path www/templates/home.html:

```
<ion-view view-title="Search">
  <ion-content class="has-header">
    <h1>A secure app!</h1>
    <div class="card">
      <div class="item item-text-wrap">
        This app contains extremely secretive confidential
        mustneverbeseen-ish information that will cause a
        disaster if it leaks out. It will also kill all
        dolphins. Please save the dolphins.
      </div>
    </div>
  </ion-content>
</ion-view>
```

The following code snippet represents the `private.html` templates at the path www/templates/private.html:

```
<ion-view view-title="Search">
  <ion-content class="has-header">
    <h1>Secret content!</h1>
    <div class="card">
      <div class="item item-text-wrap">
        You are authorized to see the grand secrets
        of the Universe!
      </div>
    </div>
  </ion-content>
</ion-view>
```

That's all that we need for the basic setup. You can verify it by running the following in a terminal or command line in the `root` folder of your directory:

```
Ionic serve -l
```

You will see the following:

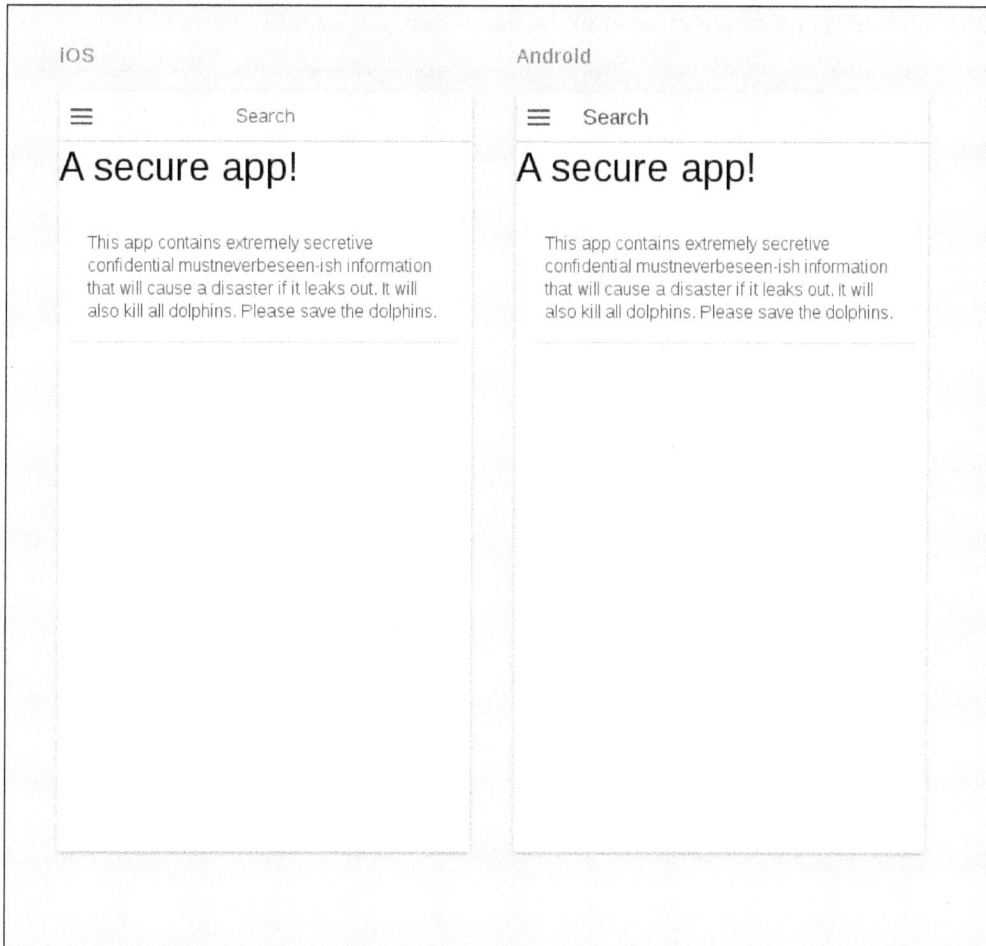

A dire warning indeed! Let's see if we can get around it. If you click on the app icon at the top left of the app screen (either for Android or iOS), you can bring out the navigation drawer that we created in the `www/templates/menu.html` file:

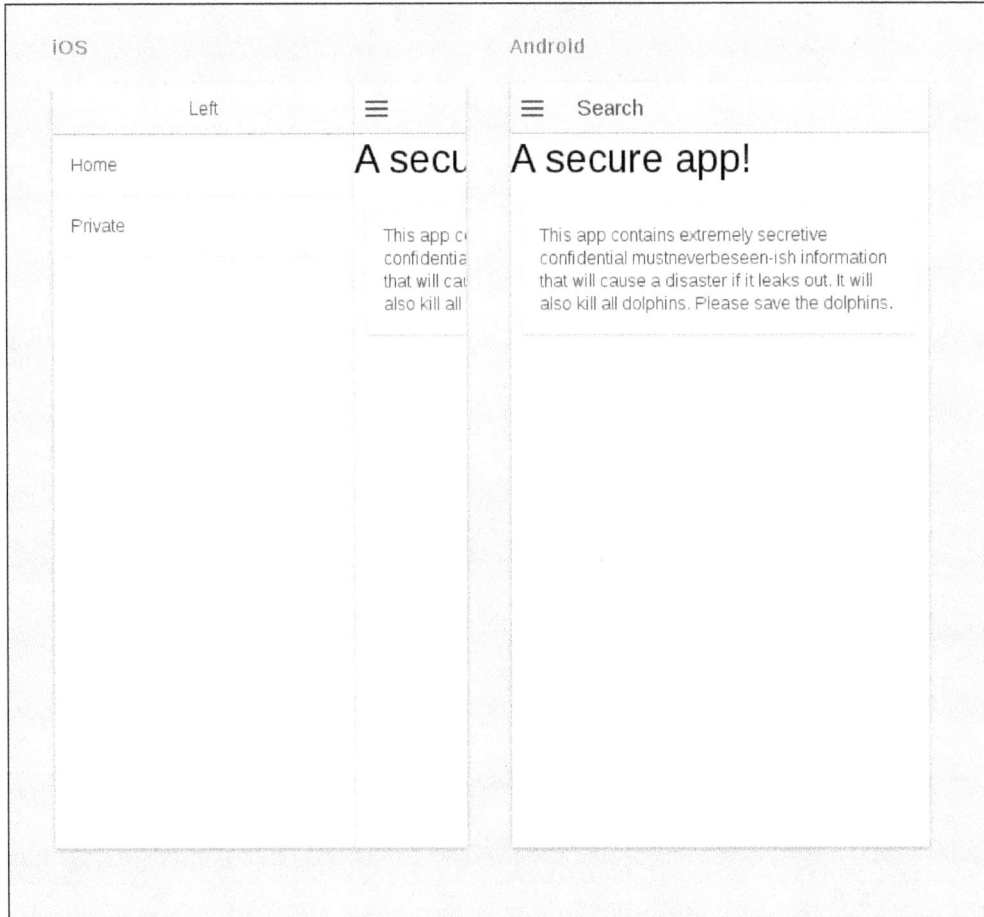

If you select the **Private** link from the list, you would expect the app to stop us from accessing information that could potentially put an end to Flipper once and for all, but alas:

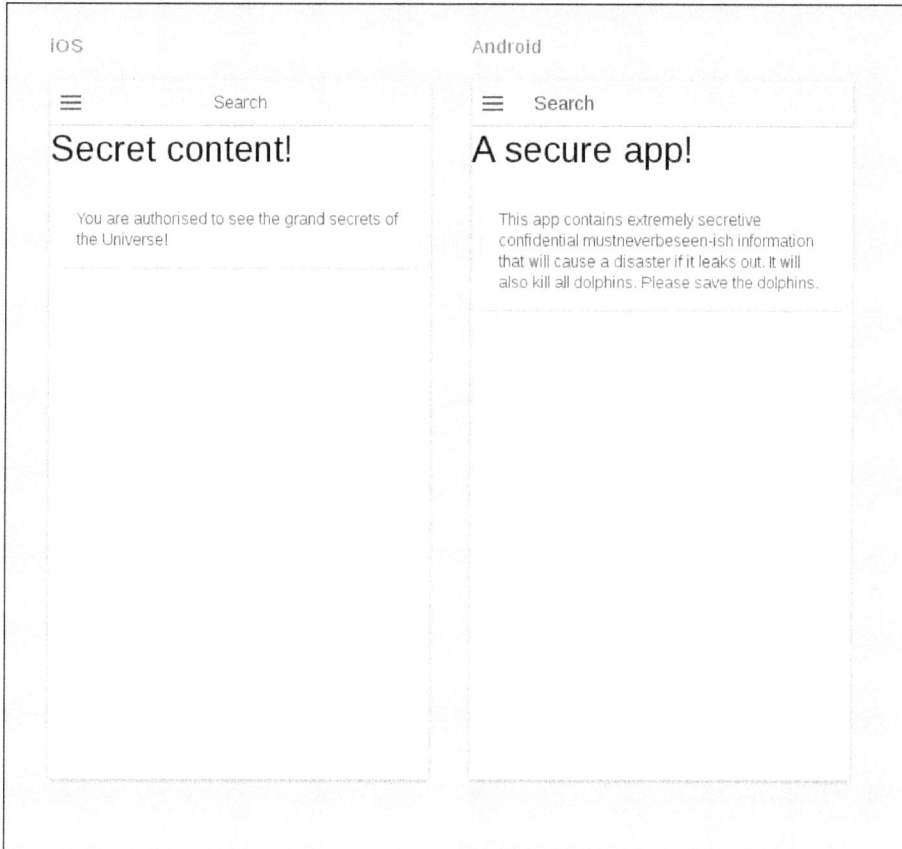

Not good! To remedy this, we will need to find a way to block the user from accessing certain content unless they are authenticated and that, even if they hack their way into accessing the content, there is no useful data for them to find anywhere.

A basic authentication service

The first step in adding basic security to our app is to create an authentication service, which can be used in order to carry out authentication requests. We want this service to provide the following functionalities:

- It should be able to log a user in. This function should take a username and password and, if they match, return an authentication token that the user can use in order to verify their identity to the system.

- It should be able to check whether a user is currently authenticated in the app. This will be necessary whenever we wish to check whether a user should have access to the system or not.

Let's go ahead and build such a service. Add the services.js file in the www/js folder and insert the following content in it:

```
angular.module('secureApp.services', [])
.factory('AuthFactory', function ($scope, $timeout) {
  var currentUser = null;
  var login = function (username, password) {
    return null;
  };
  var isAuthenticated = function () {
    return false;
  };
  var getCurrent = function () {
    return isAuthenticated() ? currentUser : null;
  };
  return {
    login: login,
    isAuthenticated: isAuthenticated,
    getCurrent: getCurrent
  }
});
```

This gives us a skeleton to work with. Let's start adding some meat to it incrementally.

The login function

The purpose of the login function is simply to take a username and password and check them against an existing list of such pairs. To get it working, we will first need to add some mock data to our service (in real life, you will of course pull the data from a remote server).

Go ahead and make sure that the `LoginFactory` contains the following:

```
var validUsers = [
{
  firstName: 'Johanna',
  lastName: 'Doe',
  username: 'johnny',
  password: 'suchsecret'
},
{
  firstName: 'Jane',
  lastName: 'Doe',
  username: 'zo1337',
  password: 'muchhide'
},
{
  firstName: 'Mary',
  lastName: 'Doe',
  username: 'bl00dy',
  password: 'wow'
}
];
```

Now, we simply need to add the following to the body of the login function:

```
var login = function (username, password) {
  var deferred = $q.defer();
  // We use timeout in order to simulate a roundtrip to a server,
  // which will be present in any realistic authentication
  scenario.
  $timeout(function () {
    // Clear any existing, cached user data before logging in
    currentUser = null;
    // See if we can find a matching username-password match
    validUsers.forEach(function (user) {
      if (user.username === username && user.password ===
      password) {
        // If we have a match, cache it as the current user
        currentUser = user;
        deferred.resolve();
      }
    });
    // If no match could be found, reject the promise
    if (!currentUser) {
```

```
        deferred.reject();
    }
}, 1000);
// Return the promise to the caller
return deferred.promise;
};
```

What we do here in terms of authentication is really quite simple. We only match usernames and passwords against a pre-defined array. If a match is found, we cache the matched user and add it to the factories context. It will now be accessible via the getCurrent() function.

The isAuthenticated function

The purpose of this function is to allow the system to check whether the current user is presently logged in or not. We can simply implement it in terms of whether there is a cached user from a successful login event available:

```
var isAuthenticated = function () {
    return currentUser ? true : false;
};
```

The getCurrent function

This function is simple, and it simply returns the current cached user:

```
var getCurrent = function () {
    return isAuthenticated() ? currentUser : null;
};
```

Implementing route authentication

Now that we have a working authentication service, let's use it in order to safeguard the world's dolphins and seal off the private part of our app. To do so, first make sure that the index.html file correctly imports the new service, as follows:

```
<script src="js/services.js"></script>
```

Next, modify the app.js file to import that file as well:

```
angular.module('secureApp',
[
  'ionic',
  'secureApp.services',
])
```

Now, in the `app.js` file, modify the routing `config` for the private part of the app so that it looks like the following code:

```
.state('app.private', {
  url: "/private",
  views: {
    'menuContent': {
    templateUrl: "templates/private.html",
    resolve: {
      isAuthenticated: function ($q, AuthFactory) {
        if (AuthFactory.isAuthenticated()) {
          return $q.when();
        } else {
          $timeout(function () {
          $state.go('app.home')
        },0);
          return $q.reject()
      }
    }
    }
});
```

What is going on here? To answer this, consider what we want to achieve. If the user is not authenticated, we want to send them back to the home screen until they have logged in. In order to do so, we perform the following steps:

1. We add a resolve hook for the transition to the `app.private` state. In terms of the router, this is a function that has to be resolved before the navigation commences.

2. Inside this hook, we use the `AuthFactory.isAuthenticated` function that we defined earlier. However, for `resolve` to work as expected, the return value of the hook needs to be a `promise` method. Thus, we use `$q` to return a when resolution if the user is logged in and a reject event if they are not.

3. If the user is not logged in, we use `$state` in order to tell the router to redirect the control to the home page again.

Finally, all we need to do is add an actual login screen for the app. To do so, start by adding a new file to keep `controllers` for our app at the path www/js/ `controllers.js`. Make sure that this file has the following content:

```
angular.module('secureApp.controllers', ['secureApp.services'])
.controller('AppCtrl', function ($scope, $ionicModal, $timeout,
AuthFactory) {
```

```
// Form data for the login modal
$scope.loginData = {};
// Create the login modal that we will use later
$ionicModal.fromTemplateUrl('templates/login.html', {
  scope: $scope
}).then(function (modal) {
  $scope.modal = modal;
});
// Triggered in the login modal to close it
$scope.closeLogin = function () {
  $scope.modal.hide();
};
// Open the login modal
$scope.login = function () {
  $scope.modal.show();
};
// Perform the login action when the user submits the login form
$scope.doLogin = function () {
  AuthFactory.login($scope.loginData.username,
  $scope.loginData.password)
  .then(function () {
    $scope.closeLogin();
  });
};
});
```

To render the login screen itself, add a template for to the path www/templates/
login.html:

```
<ion-modal-view>
  <ion-header-bar>
    <h1 class="title">Login</h1>
    <div class="buttons">
      <button class="button button-clear"
      ng-click="closeLogin()">
        Close
      </button>
    </div>
  </ion-header-bar>
  <ion-content>
    <form ng-submit="doLogin()">
      <div class="list">
        <label class="item item-input">
          <span class="input-label">
            Username
```

```
        </span>
        <input type="text"
        ng-model="loginData.username">
      </label>
      <label class="item item-input">
        <span class="input-label">
          Password
        </span>
        <input type="password"
        ng-model="loginData.password">
      </label>
      <label class="item">
        <button class="button button-block button-positive"
        type="submit">
          Log in
        </button>
      </label>
    </div>
  </form>
</ion-content>
</ion-modal-view>
```

Finally, let's tie everything together by making sure that the app loads our newly defined controller. Load it in `index.html`:

```
<script src="js/controllers.js"></script>
```

Next, make sure that it is listed as a dependency in `app.js`:

```
angular.module('secureApp',
[
  'ionic',
  'secureApp.services',
  'secureApp.controllers'
])
```

We are now building our app. You can try it out by running it yourself. Try logging in with wrong credentials (according to the ones that we defined) in order to convince yourself that the app really blocks the user from going where they should not.

Summary

In this chapter, you gained a basic understanding of how client-side authentication works and what its basic limitations are. You also saw how to create a basic app that implements some of the basic concepts to create an app in order to see how the app works in practice.

In the next chapter, you'll learn how to set up web socket communication through the app in order to subscribe to dynamic notifications from a server.

12
Working with Real-Time Data

In today's app ecosystem, real-time features of various sorts are more or less becoming staple. Chat applications (and chat features for existing apps) are ever more common, push notifications bring news and views to users without them having to look for it, and so on.

In this chapter, we will take a look at how we can incorporate some choice real-time features into our Ionic apps. We will do so by building a simple chat application without authentication, where at least two people can get together and talk about the wonders of life. In doing so, we will revisit what we learnt earlier about web sockets, as we will need to create a simple server for this end.

A refresher – WebSockets

Before we move on, let's have a quick refresher on an important concept that we visited earlier — WebSockets.

WebSockets is a standardized Internet protocol, which allows for direct server-to-client communication over a network. This is rather unusual in the world of traditional client-server architecture, where almost all communication is initiated by the client and the server simply responds to such communication.

WebSockets makes it easy to build real-time apps because the server can dynamically push new data to the connected clients as soon as its state changes. This is ideal for chat applications, in which we would otherwise have to use more tedious and resource-consuming approaches, such as polling, in order to approximate the same effect.

For a deeper understanding of WebSockets, please refer to *Chapter 5, Real-Time Data and WebSockets*, where we dealt with them in depth.

Getting the lay of the land

In this chapter, we are going to build a chat application that is hauntingly reminiscent of the one that we saw in *Chapter 5, Real-Time Data and WebSockets*. The big difference, of course, is that our client will be an Ionic app this time, which will be able to interact fully with browser clients that are also connected to the same server. In doing so, we demonstrate how easy it is to build apps that almost seamlessly interact with apps on other platforms that use the same server.

What we will need

To get our app working, we will need:

- A server that can both receive and relay messages via WebSockets.
- An app that can connect to a server using WebSockets and send, receive, and process messages over the same protocol. All the sent data should be rendered in a way that is meaningful to the user.

In the spirit of this book, we will of course use Node.js for our server. To add WebSocket support to it, we will use the socket.io library, which you already saw in *Chapter 6, Introducing Ionic*.

For the client, we will use the standard socket.io client library, which was seen in *Chapter 5, Real-Time Data and WebSockets*, (observed a trend here?). However, we will make some clever use of it in order to make things work smoothly on the app side of things.

Let's go ahead and start setting things up!

Creating the server

The first thing that we need to do is create a WebSocket server to relay messages between our clients. Find a suitable project folder, open your terminal/command line, and run the following:

```
npm init
```

This will create the basic Node.js project structure. You can enter whatever values you see fit:

```
{
  "name": "ionic-chat-server",
  "version": "1.0.0",
```

```
    "description": "A websocket server for chatting.",
    "main": "server.js",
    "scripts": {
      "test": "echo \"Error: no test specified\" && exit 1"
    },
    "author": "csvan",
    "license": "MIT"
  }
```

Now, let's install the dependencies that we will need. Run the following command:

```
npm install socket.io
```

This will install `socket.io`, which is all we will need in order to get our server running.

Next, create the `server.js` file in the current folder and add the following content to it:

```
var http = require('http');
var url = require('url');
var fs = require('fs');
var server = http.createServer(function (req, res) {
  var parsedUrl = url.parse(req.url, true);
  switch (parsedUrl.pathname) {
    case '/':
    // Read the file into memory and push it to the client
    fs.readFile('index.html', function (err, content) {
      if (err) {
        res.writeHead(500);
        res.end();
      }
      else {
        res.writeHead(200, {'Content-Type': 'text/html'});
        res.end(content, 'utf-8');
      }
    });
    break;
  }
});
// Connect the websocket handler to our server
var websocket = require('socket.io')(server);

// Create a handler for incoming websocket connections
```

```
websocket.on('UserConnectedEvent', function (socket) {
  console.log("New user connected");

  // Tell others a new user connected
  socket.broadcast.emit('UserConnectedEvent', null);

  // Bind event handler for incoming messages
  socket.on('MessageSentEvent', function (chatData) {
    console.log('Received new chat message');

    // By using the 'broadcast' connector, we will
    // send the message to everyone except the sender.
    socket.broadcast.emit('MessageReceivedEvent', chatData);
  });
});
server.listen(8080);
```

Looks familiar, doesn't it? This is pretty much the exact same server that we developed back in *Chapter 5, Real-Time Data and WebSockets*! It even has that beautiful, modern chat UI that we built there. All that we need to do is serve it. To do so, add the index.html file to the current folder and add the following to it:

```
<!DOCTYPE html>
<html>
<head lang="en">
  <meta charset="UTF-8">
    <title>Socket.io chat application</title>
    <link rel="stylesheet" href="http://maxcdn.bootstrapcdn.com/
    bootstrap/3.3.4/css/bootstrap.min.css"/>
</head>
  <body>
    <ul id="messages"></ul>
    <div class="container">
      <div class="row">
        <div id="chat-box" class="well">
          <ul id="chat-view" class="list-unstyled"></ul>
        </div>
      </div>
      <form action="">
        <div class="row">
        <input type="text"
        id="chat-name"
```

```
      class="form-control"
      placeholder="Your name">
      </div>
      <div class="row">
        <input type="text"
        id="chat-message"
        class="form-control"
        placeholder="Enter message">
        <button id="chat-submit"
        type="submit"
        class="btn btn-default">Send
        </button>
      </div>
    </form>
</div>
<script src="/socket.io/socket.io.js"></script>
<script src="https://code.jquery.com/jquery-
1.11.0.min.js"></script>
<script>
  var websocket = io();
  var appendChatMessage = function (sender, message) {
    $('#chat-view').append($('<li>').text(sender + ': ' +
    message));
  };
  var clearChatField = function () {
    $('#chat-message').text('');
  };
  // Notify the server when we send a new message

  $('#chat-submit').click(function () {
    var chatData = {
      name: $('#chat-name').val(),
      message: $('#chat-message').val()
    };
    appendChatMessage(chatData.name, chatData.message);
    clearChatField();
    websocket.emit('newChatMessage', chatData);
    return false;
```

```
    });

    // Update the state of the chat when we receive a new chat
    message
    websocket.on('new chat message', function (chatData) {
      appendChatMessage(chatData.name, chatData.message);
    });
  </script>
 </body>
</html>
```

That's all there is to it. Our browser-based chat is now all up and running...again! I won't explain how everything works here. Checkout *Chapter 5, Real-Time Data and WebSockets*, if anything seems unclear. Whenever you are ready, let's head right on and get started with our app client.

Building the chat app

One of our primary concerns when developing mobile experiences is creating an interface that is intuitive for users to use. Fortunately, Ionic comes loaded with some very convenient features to make this possible. Before we get there though, let's set up the basics.

Setting up the basic app structure

Let's start by creating a basic app. Create a suitable project folder, which is different from that of the chat server that we created earlier. Navigate to the folder, open a terminal/command line, and run the following:

```
ionic start ionic-chat-app blank
```

As we have seen before, this will create an empty Ionic project for us to fill with love and good things. Have a look around. Not much to see here, right? We will change that soon enough. Hang tight.

It's early, but let's get our Ionic preview server going right away so that we can see the app live. Without changing anything, run the following from your terminal / command line:

```
ionic serve -l
```

This will bring up the preview for both the Android and iOS displays. As expected, there is not much to see here yet:

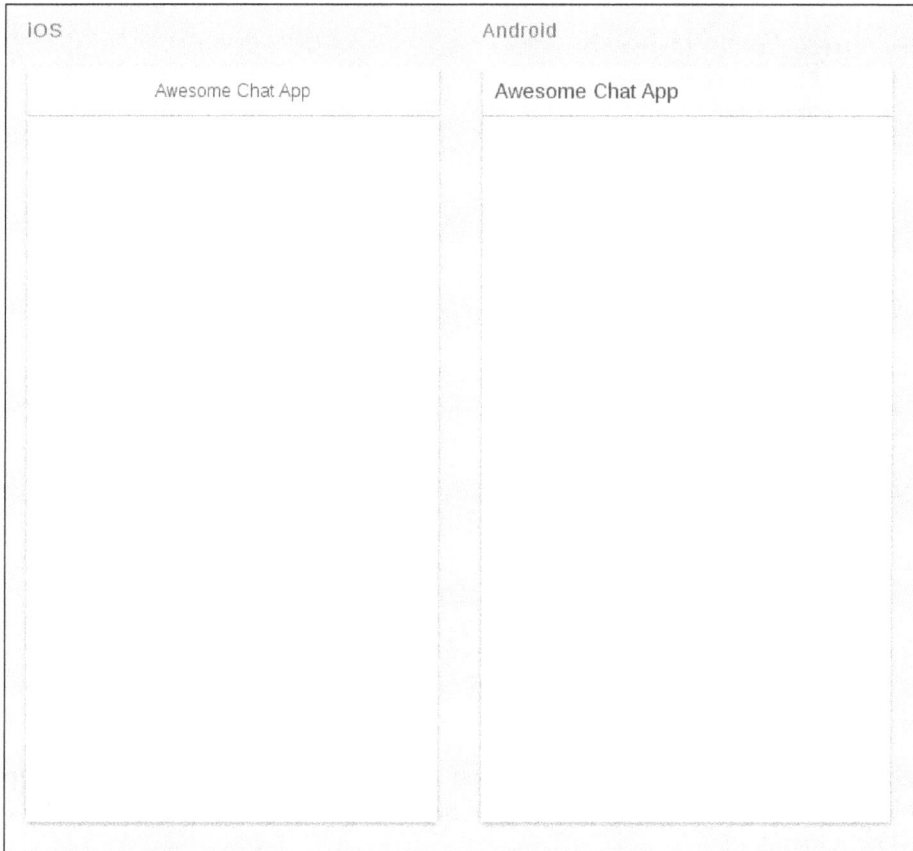

Let's go ahead and set up the basics. The first thing that we need to deal with is routing. In the js/app.js file, make sure that you have the following:

```
angular.module('ionic-chat-app', ['ionic'])
.run(function ($ionicPlatform) {
  $ionicPlatform.ready(function () {
    if (window.cordova && window.cordova.plugins.Keyboard) {
      cordova.plugins.Keyboard.hideKeyboardAccessoryBar(true);
    }
    if (window.StatusBar) {
```

```
        StatusBar.styleDefault();
      }
    })
  })
  .config(function ($stateProvider, $urlRouterProvider) {
    // Configure the routing
    $stateProvider.
    state('app', {
      url: "/app",
      abstract: true,
      templateUrl: 'index.html'
    })
    .state('app.chat', {
      url: '/chat',
      templateUrl: 'templates/app-chat.html'
    });
    $urlRouterProvider.otherwise('/app/chat');
  });
```

Here, we defined a basic, abstract state called app, which we will leave as the root state for the app as a whole. The only child state of this state is app.state, which will contain the actual chat view and associated logic.

It may seem counter-intuitive to add states for an app that we really only want to have a single view of. However, this is a good architectural precaution to take in case we want to expand the app further.

Now, let's add some basic view information. Create a **templates** folder in your app's www folder and then proceed to create a file named app-chat.html in it, which has the following content:

```
<ion-view view-title="chat">
  <ion-content class="padding">
    Chat awesomeness goes here!
  </ion-content>
</ion-view>
```

Now, your app preview should look like this:

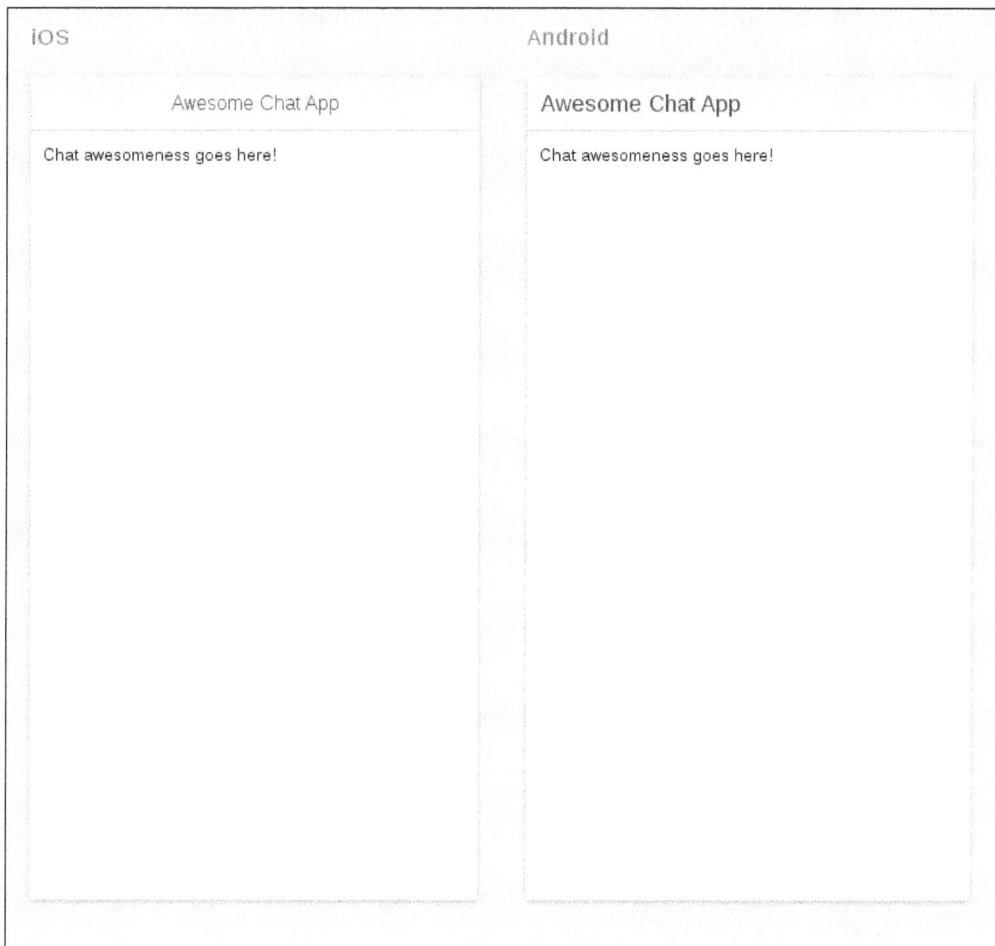

This is slightly better, but we're not quite there yet. Next, we will add the actual chat layout to it.

The input section

Go ahead and modify the `app-chat.html` file so that it looks like this:

```
<ion-view view-title="chat">
  <ion-content class="padding">
  </ion-content>
  <div class="bar bar-footer bar-balanced">
    <label class="item-input-wrapper">
      <input id="message-input" type="text" placeholder="Message">
    </label>
    <button class="button button-small">
      Submit
    </button>
  </div>
</ion-view>
```

Here, we attached a footer to our app—an element that will be permanently fixed to the bottom of the viewport. Inside this footer, we defined an input field to add a message and an associated button to actually send it. To make the input box scale appropriately, we need to add the following to the `css/style.css` file:

```
#message-input {
  width: 100%;
}
```

Having done all this, we will end up with the following:

So far, so good. *Gotta* love that spicy green touch. On we go!

The message view

Now, let's create the part of our app that will display all the messages in our most important chat.

Modify the `templates/app-chat.html` file so that the `<ion-content>` tag looks like the following:

```
<ion-content>
  <div class="list">
    <a class="item item-avatar">
      <h2>Me</h2>
      <p>Anyone out there!?</p>
    </a>
    <a class="item item-avatar other-chatbox">
      <h2>Anyone</h2>
      <p>Yes.</p>
    </a>
    <a class="item item-avatar">
      <h2>Me</h2>
      <p><3</p>
    </a>
  </div>
</ion-content>
```

The preceding code creates a standard Ionic list that contains a set of `item-avatar` elements. These are standard list items in Ionic, which make it easy to show an avatar image, a heading, and some text, as shown in the following example from the Ionic element reference:

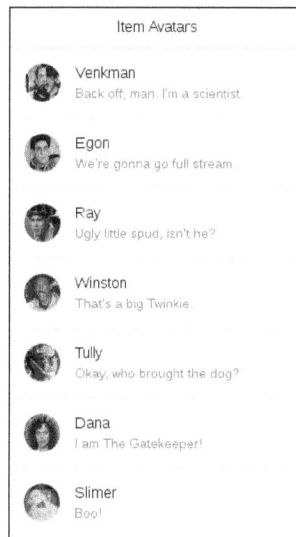

However, in our case, we will skip over the actual images and just use the header and text. These two make for a very convenient way of showing a single chat message along with the name of the person sending it.

Next, add the following to the `css/style.css` file:

```css
.item-avatar {
  padding-left: 16px;
}
.other-chatbox {
  text-align: right;
}
```

The preceding code is needed in order to override the default Ionic styling for the `item-avatar` element. This allows for the title and text to be positioned either right or left. This will bring the final look closer to the more popular chat apps, where the texts of other participants are usually positioned to the right of the flow, whereas our own are positioned to the left.

Your preview should now look like this:

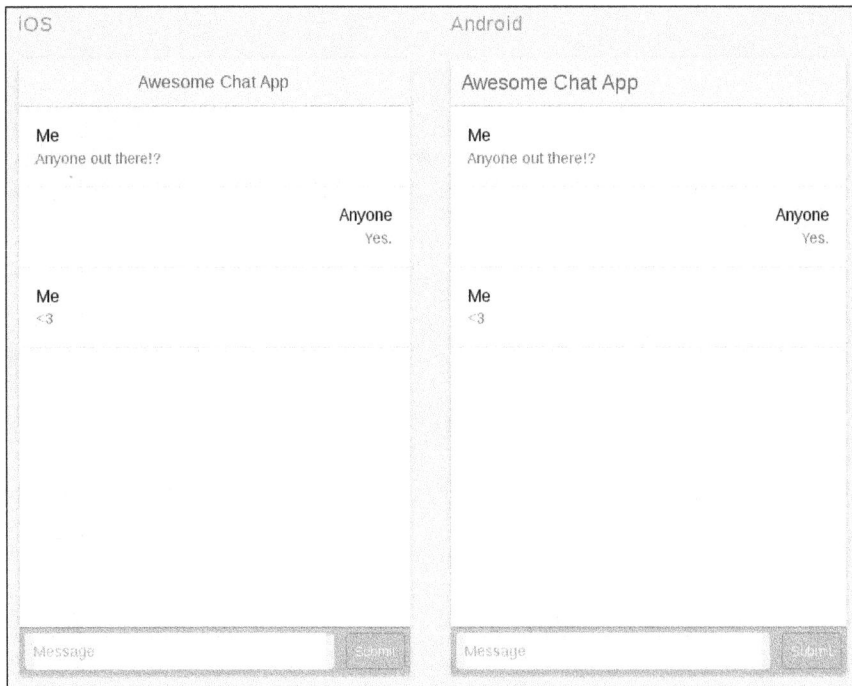

That does it for a very basic chat interface. Now, let's go ahead and add some logic to it all.

The ChatService function

In order to communicate with the WebSocket server, we will create a service that will allow us both to send messages to the server as well as subscribe to the events sent by it.

To start off, create a file named `js/services.js` and insert the following code in it:

```
angular.module('ionic-chat-app-services', [])
.service('ChatService', function ChatService($rootScope) {
  this.emit = function (message) {
    // Send a message
  };
  this.on = {
    userConnected: function (callback) {
      $rootScope.$on('UserConnectedEvent',
      function (event, user) {
        callback(user);
      })
    },
    messageReceived: function (callback) {
      $rootScope.$on('MessageReceivedEvent',
      function (event, message) {
        callback(message);
      })
    }
  }
});
```

Our `service` here exposes the following two core features to the user:

- `Emit`: This allows the user to broadcast a message to the server
- `On`: This allows the user to subscribe to the following two events:

    - `UserConnectedEvent`: This is fired whenever a new user connects to the app
    - `MessageReceivedEvent`: This is fired whenever a new message is received from the server

The passing of messages in the preceding code is implemented by means of the `$rootScope` function, which already provides us with a robust mechanism. We merely wrap it up in order to meet our own ends.

To integrate `service` into our app, add the following to the `index.html` file:

```
<script src="js/app.services.js"></script>
```

Next, list the chat services as a dependency in the `js/app.js` file, as follows:

```
angular.module('ionic-chat-app',
[
  'ionic',
  'ionic-chat-app-services'
])
```

Adding WebSockets to the mix

Now that the service is connected to the app, let's bring WebSockets into play to actually make it do something fun! To start off, add the following to the `index.html` file in order to import the `socket.io` client library:

```
<script src="https://cdn.socket.io/socket.io-x.x.x.js"></script>
```

Replace the `x.x.x` in the preceding code with whatever version of `socket.io` you are running on your server (if you are not sure, check the `package.json` file in the chat server project that we created earlier).

This will give us the global object named `io`, which can be used to interact with a WebSocket server. Global objects are evil. Very evil. So, as a matter of good principle, we will make the best effort to contain it in our chat service, as follows:

```
angular.module('ionic-chat-app-services', [])
.service('ChatService', function ChatService($rootScope) {

  // Init the Websocket connection
  var socket = io.connect('http://localhost:8080');

  // Bridge events from the Websocket connection to the rootScope
  socket.on('UserConnectedEvent', function(user) {
    $rootScope.emit('UserConnectedEvent', user);
  });
  socket.on('MessageReceivedEvent', function(message) {
    $rootScope.emit('MessageReceivedEvent', message);
  });

  /*
  * Send a message to the server.
  * @param message
  */
  this.emit = function (message) {
```

```
    socket.emit('MessageSentEvent', message);
  };
  this.on = {
    userConnected: function (callback) {
      $rootScope.$on('UserConnectedEvent',
      function (event, user) {
        callback(user);
      })
    },
    messageReceived: function (callback) {
      $rootScope.$on('MessageReceivedEvent',
      function (event, message) {
        callback(message);
      })
    }
  }
});
```

What we have done here is pretty straightforward and can be summarized as follows:

- We listen for events from the WebSocket server using the on function and simply pass these events along to the $rootScope function. By doing so, the other parts of our app can register listeners and callbacks for these events in order to act on them.

- We use the emit function of the socket in order to send messages back to the server.

This concludes the hard logic behind our app. Next, we will tie it all together by making our chat view dynamic.

Updating the chat view

Whenever you or another connected user submits something to the chat, you want the chat display to show the new message. If that sounds like a job for ng-repeat, that's because it...well...really isn't.

While ng-repeat is a very powerful directive on its own, it can unfortunately incur severe performance penalties as data sets grow over time, especially when it comes to mobile devices, where processing power is limited. To work around this, Ionic offers another directive to render dynamic datasets—collection-repeat. Without much intervention from our side, collection-repeat will do a lot of the really heavy lifting when it comes to working with collections.

However, before we can do this, we will need to add a controller for our chat view. Go ahead and create the `js/app.controllers.js` file. Import it and add it as a dependency in `index.html` and `app.js` respectively:

```
<script src="js/app.controllers.js"></script>

angular.module('ionic-chat-app',
[
  'ionic',
  'ionic-chat-app-services',
  'ionic-chat-app-controllers'
])
```

Next, let's add some basic content to the file:

```
angular.module('ionic-chat-app-controllers', [])
.controller('ChatController', function ($scope) {
});
```

Finally, let's bind the controller to our app's chat state. In `app.js`, make sure that your state definition looks like the following:

```
$stateProvider.
state('app', {
  url: "/app",
  abstract: true,
  templateUrl: 'index.html'
})
.state('app.chat', {
  url: '/chat',
  templateUrl: 'templates/app-chat.html',
  controller: 'ChatController'
});
```

We are now ready to start adding some serious functionality to our app! Go ahead and add the following to your controller:

```
angular.module('ionic-chat-app-controllers', [])
.controller('ChatController', function ($scope, ChatService) {
  // The chat messages
  $scope.messages = [];
  // Notify whenever a new user connects
  ChatService.on.userConnected(function (user) {
    $scope.messages.push({
      name: 'Chat Bot',
```

```
      text: 'A new user has connected!'
    });
  });
  // Whenever a new message appears, append it
  ChatService.on.messageReceived(function (message) {
    message.external = true;
    $scope.messages.push(message);
  });
  $scope.inputMessage = '';
  $scope.onSend = function () {
    $scope.messages.push({
      name: 'Me',
      text: $scope.inputMessage
    });
    // Send the message to the server
    ChatService.emit({
      name: 'Anonymous',
      text: $scope.inputMessage
    });
    // Clear the chatbox
    $scope.inputMessage = '';
  }
});
```

Finally, modify the `templates/app-chat.html` file in order to connect to the controller data, as follows:

```
<ion-view view-title="chat">
  <ion-content>
    <div class="list">
      <a collection-repeat="message in messages"
      class="item item-avatar"
      ng-class="{'other-chatbox' : message.external}">
        <h2>{{message.name}}</h2>
        <p>{{message.text}}</p>
      </a>
    </div>
  </ion-content>
  <div class="bar bar-footer bar-balanced">
    <label class="item-input-wrapper">
      <input id="message-input"
      type="text"
      placeholder="Message"
      ng-model="inputMessage">
```

```
      </label>
      <button class="button button-small"
      ng-click="onSend()">
        Submit
      </button>
    </div>
  </ion-view>
```

Note that we are bringing the fabled `collection-repeat` function into play here with pretty much no configuration needed! The app will now respond both to our own message sending events as well as others' messages that are coming in from the server. Fire it up in your preview and try it out!

Going further

Here, to keep things simple, we simulated our chat app on the emulator. However, we could have of course had even more fun if we actually got it running on a set of physical devices. If you are the kind of person with so much money that you can pick up a bunch of iPhones and Android devices on your way back from the grocery store (or just have a lot of friends with the same devices), why not make it a fun project of hosting your chat server on an actual **VPS (Virtual Private Server)** and connect the project to it? You and your friends can discuss plans for world domination in your very own app!

The VPS that you want to use is up to you, but we can think of several options for you to consider. Check out `https://www.digitalocean.com/` and `https://www.linode.com/` to fire up your backend. There are also several more specific solutions that enable you to fire up your backend with less configuration such as **Heroku**. Many of the prominent services make it extremely easy to configure the more crucial elements of server functionality such as DNS.

Once you configure your VPS and run your chat server, change the target domain in your app from localhost to the domain/IP of your server. Invite your friends and chat away! However, in order to make sure that the integrity of your users is protected, make sure that you enable communication only over HTTPS for actual live applications.

Summary

In this chapter, we explored how we can incorporate real-time functionality into an Ionic app using `socket.io`. In the process of doing so, we built a simple chat application, which can be expanded later in order to learn advanced real-time features.

In the next chapter, we'll take a look at how to set up WebSocket communication through the app in order to subscribe to dynamic notifications from a server. We will elaborate on how this helps us develop truly dynamic applications such as chat apps.

13
Building an Advanced Chat App

In the previous chapter, we developed a rudimentary chat application, which allowed an arbitrary number of users to connect to each other and talk anonymously.

In this chapter, we are going to expand this app and make it more advanced by adding features for chat rooms and notifications. In doing so, we will demonstrate how the concept of namespacing works on `socket.io`, which is one of the most important aspects of this library.

We need some room!

So far, the most advanced thing that we have done with WebSockets in our apps has simply been sending data back and forth across a single WebSocket interface. We paid very little attention to partitioning and basically just let it all go on as a free-for-all app. However, in real life, we will frequently find ourselves in situations where we want to partition WebSocket connections and only let certain users have access to a subset of partitions.

To see how this can work, consider the case of a *group chat*. Here, rather than having just a single solitary chat interface, users instead have access to a multitude of them; each hosts its own members and conversation. To implement this, we can extend our existing chat server to simply start new `node` instances for the chat rooms that we want to open, with each of them having its own port, as follows:

```
// [snip]

// Connect the websocket handler to our server
var websocket = require('socket.io')(server);
```

```
// Create a handler for incoming websocket connections
websocket.on('UserConnectedEvent', function (socket) {
  console.log("New user connected");
  // Tell others a new user connected
  socket.broadcast.emit('UserConnectedEvent', null);
  // Bind event handler for incoming messages
  socket.on('MessageSentEvent', function (chatData) {
    console.log('Received new chat message');
    // By using the 'broadcast' connector, we will
    // send the message to everyone except the sender.
    socket.broadcast.emit('MessageReceivedEvent', chatData);
  });
});

// Define a separate port for each server we want to start
var port = 8080; // get from terminal args, for example

server.listen(port);
```

However, this becomes clunky very quickly. Since we will need to fire up a new, separate V8 instance for each server, chances are that we will very soon get angry knocks at the office window from the guy down the hall whose super-important stock analysis algorithm just crashed due to a lack of memory space. He may want to hurt us and do terrible things to our pets, all because we could not find a smoother way to make use of WebSockets.

Or, well, maybe we can, after all.

This is where the concept of a namespace comes into play. Imagine a situation where we can just partition a single socket.io instance into several different endpoints, each of which can service its own set of clients. It turns out that we can!

Namespaces

Remember how we set up our original websocket server? For that take a look at the following code:

```
// [snip]

// Connect the websocket handler to our server
var websocket = require('socket.io')(server);
// Create a handler for incoming websocket connections
websocket.on('UserConnectedEvent', function (socket) {
  console.log("New user connected");
```

```
      // Tell others a new user connected
      socket.broadcast.emit('UserConnectedEvent', null);
      // Bind event handler for incoming messages
      socket.on('MessageSentEvent', function (chatData) {
        console.log('Received new chat message');
        // By using the 'broadcast' connector, we will
        // send the message to everyone except the sender.
        socket.broadcast.emit('MessageReceivedEvent', chatData);
      });
    });
```

Here, we simply used the main socket.io instance in order to directly register socket connections and their respective callbacks. However, looking a little closer, what we actually did was connect all the incoming connections to a namespace, even if this happened implicitly. Take a look at the following snippet from the preceding code:

```
    websocket.on('UserConnectedEvent', function (socket) {
      console.log("New user connected");
      // Tell others a new user connected
      socket.broadcast.emit('UserConnectedEvent', null);
      // Bind event handler for incoming messages
      socket.on('MessageSentEvent', function (chatData) {
        console.log('Received new chat message');
        // By using the 'broadcast' connector, we will
        // send the message to everyone except the sender.
        socket.broadcast.emit('MessageReceivedEvent', chatData);
      });
```

What is actually happening here is that we are registering the connections on the root namespace (written as /), which is the one namespace that socket.io gives us to work with even if we specify no other namespaces. This goes to show that namespaces are actually essential for the way socket.io works internally. In fact, every single connection that you have going will be associated with a single namespace, even if it is an implicit one!

Your users connect to the root namespace whenever they connect directly to the URL of your WebSocket server. For example, they can do this by doing the following on the client side:

```
    var socket = io.connect('http://localhost:8080');
```

You are in effect telling socket.io that you wish to establish a connection to the root namespace.

The problem is that if we perform the preceding steps (as we have done until now), all the messages that we send to the server are open for broadcasting to all the other connected clients as well (this happens even if you have other namespaces defined, as we will see later). This is not very convenient if we want to concentrate on communications.

Let's say that we want to divide communications in order to let users subscribe to websocket channels, which sends information that interests them. For example, let's say that we are building a chat application that will let them speak about various programming languages such as Java, Scala, and JavaScript. In that case, we can define namespaces on the server side by doing the following:

```
websocket.of('/java').on('UserConnectedEvent', function (socket) {
  console.log("New user connected to the Java channel");
  socket.broadcast.emit('UserConnectedEvent', null);
  socket.on('MessageSentEvent', function (chatData) {
    console.log('Received new Java chat message');
    socket.broadcast.emit('MessageReceivedEvent', chatData);
});

websocket.of('/scala').on('UserConnectedEvent',
function (socket) {
  console.log("New user connected to the Scala channel");
  socket.broadcast.emit('UserConnectedEvent', null);
  socket.on('MessageSentEvent', function (chatData) {
    console.log('Received new Scala chat message');
    socket.broadcast.emit('MessageReceivedEvent', chatData);
});

websocket.of('/javascript').on('UserConnectedEvent',
function (socket) {
  console.log("New user connected to the Java channel");
  socket.broadcast.emit('UserConnectedEvent', null);
  socket.on('MessageSentEvent', function (chatData) {
    console.log('Received new Javascript chat message');
    socket.broadcast.emit('MessageReceivedEvent', chatData);
});
```

The important parts of the code are emphasized. Note how we use the `of` function in order to create the actual namespace. The argument of the function is the name of the namespace relative to the root namespace (/).

After the namespace is created, we register `socket` connections in a way that is familiar to us by now after having (albeit unknowingly!) done the same thing with the root namespace earlier.

We can now make use of these modifications to the server by having the client connect to any given namespace available. For example, for the ones that we have already defined here, you can connect to each of them like this (respectively):

```
var javaSocket = io.connect('http://localhost:8080/java');

var scalaSocket = io.connect('http://localhost:8080/scala');

var javascriptSocket = io.connect('http://localhost:8080/javascript');
```

Then, proceed to operate on them just as you would in the case of any other single WebSocket connection, as follows:

```
javaSocket.on('UserConnectedEvent', function (user) {
  console.log('User connected to the Java channel:', user);
});
```

This is all pretty straightforward, as you will note as you dig in a little deeper. Let's do so by dusting off the simple chat application that we wrote in the last chapter and giving it some genuine namespacing love.

Creating a multiroom chat application

Let's take a brief refresher on the basic **Chat App** that we built during the course of the previous chapter:

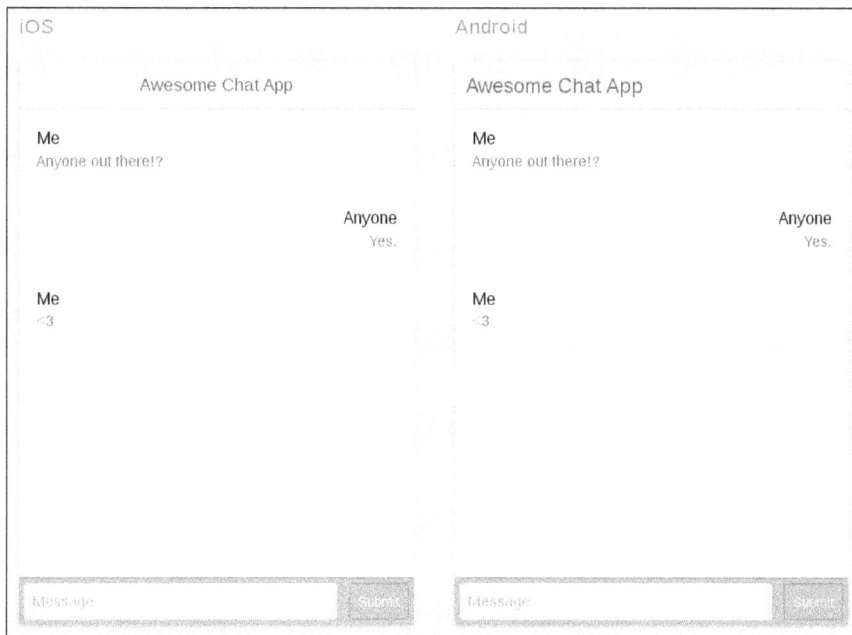

This app effectively sets up a connection to the WebSocket server and lets us talk to random strangers who, for some reason, are loitering in the kitchen and using the Wi-Fi connection. What we want to do here is give these strangers (and ourselves) the possibility to pick separate chat rooms depending on what they are keen to talk about. Since we love programming, programming languages are of course going to be the be-all-and-end-all of what is on the menu.

Configuring the basic layout

In order to create a nice way to navigate between different chat rooms, we will use a tabbed layout, where each tab will correspond to a single chat room.

This means that we will need to make several changes to our HTML as well as the routing for our app. Start out by modifying the `index.html` file. Make sure that it looks like the following:

```
<!DOCTYPE html>
<html>
  <head>
    <meta charset="utf-8">
    <meta name="viewport" content="initial-scale=1,
    maximum-scale=1, user-scalable=no, width=device-width">
    <title></title>
    <link href="lib/ionic/css/ionic.css" rel="stylesheet">
    <link href="css/style.css" rel="stylesheet">
    <!-- IF using Sass (run gulp sass first), then uncomment
    below and remove the CSS includes above
    <link href="css/ionic.app.css" rel="stylesheet">
    -->
    <!-- ionic/angularjs js -->
    <script src="lib/ionic/js/ionic.bundle.js"></script>
    <!-- cordova script (this will be a 404 during development)
    -->
    <script src="cordova.js"></script>
    <!-- your app's js -->
    <script src="https://cdn.socket.io/socket.
    io-1.3.5.js"></script>
    <script src="js/app.services.js"></script>
    <script src="js/app.controllers.js"></script>
    <script src="js/app.directives.js"></script>
    <script src="js/app.js"></script>
```

```
    </head>
    <body ng-app="ionic-chat-app">
      <ion-nav-bar class="bar-stable">
        <ion-nav-back-button>
        </ion-nav-back-button>
      </ion-nav-bar>
      <ion-nav-view></ion-nav-view>
    </body>
</html>
```

I highlighted the most important part in the preceding code. Here, we created a navigation bar, which corresponds to a toolbar at the top of the screen in Ionic. If you are familiar with Android, you will recognize this as the action bar. Below this navigation bar, we then attached the actual view, which is currently loaded.

Next, we will attach a series of tabs to this layout, which will let us select the chat room that we wish to interact with. In the templates folder, create a file named tabs.html and make sure that it has the following content:

```
<ion-tabs class="tabs-icon-top tabs-color-active-positive">
  <!-- Node chat -->
  <ion-tab title="Node Chat"
  icon-off="ion-ios-chatboxes-outline"
  icon-on="ion-ios-chatboxes"
  href="#/app/node">
    <ion-nav-view name="node-view">
    </ion-nav-view>
  </ion-tab>
  <!-- Javascript chat -->
  <ion-tab title="JS Chat"
  icon-off="ion-ios-chatboxes-outline"
  icon-on="ion-ios-chatboxes"
  href="#/app/javascript">
    <ion-nav-view name="javascript-view">
    </ion-nav-view>
  </ion-tab>
  <!-- Haskell chat -->
  <ion-tab title="Haskell Chat"
  icon-off="ion-ios-chatboxes-outline"
  icon-on="ion-ios-chatboxes"
  href="#/app/haskell">
    <ion-nav-view name="haskell-view">
```

```
        </ion-nav-view>
      </ion-tab>
      <!-- Erlang chat -->
      <ion-tab title="Erlang Chat"
      icon-off="ion-ios-chatboxes-outline"
      icon-on="ion-ios-chatboxes"
      href="#/app/erlang">
        <ion-nav-view name="erlang-view">
        </ion-nav-view>
      </ion-tab>
      <!-- Scala chat -->
      <ion-tab title="Scala Chat"
      icon-off="ion-ios-chatboxes-outline"
      icon-on="ion-ios-chatboxes"
      href="#/app/scala">
        <ion-nav-view name="scala-view">
        </ion-nav-view>
      </ion-tab>
    </ion-tabs>
```

Here, we used the `ion-tabs` directive, which in essence acts like a horizontal list consisting of `ion-tab` instances. Note how we associate each tab with a single language view and URL. The router will use both in order to deduce the exact state the app should be in when a tab is clicked. Let's see how it does so. Open the `app.js` file and make sure that it looks like the following:

```
angular.module('ionic-chat-app',
[
  'ionic',
  'ionic-chat-app-services',
  'ionic-chat-app-controllers'
])
.run(function ($ionicPlatform) {
  $ionicPlatform.ready(function () {
  if (window.cordova && window.cordova.plugins.Keyboard) {
    cordova.plugins.Keyboard.hideKeyboardAccessoryBar(true);
  }
  if (window.StatusBar) {
    StatusBar.styleDefault();
  }
})
})
```

```
.config(function ($stateProvider, $urlRouterProvider) {
  // Configure the routing
  $stateProvider
  // Each tab has its own nav history stack:
  .state('app', {
     url: '/app',
     abstract: true,
     templateUrl: "templates/tabs.html"
  })
  .state('app.node', {
  url: '/node',
  views: {
     'node-view': {
       templateUrl: 'templates/app-chat.html',
       controller: 'ChatController',
       resolve: {
         chatRoom: function () {
           return 'node';
         }
       }
     }
   }
 }
})
.state('app.javascript', {
  url: '/javascript',
  views: {
     'javascript-view': {
       templateUrl: 'templates/app-chat.html',
       controller: 'ChatController',
       resolve: {
         chatRoom: function () {
           return 'javascript';
         }
       }
     }
   }
})
.state('app.haskell', {
  url: '/haskell',
  views: {
     'haskell-view': {
```

```
        templateUrl: 'templates/app-chat.html',
        controller: 'ChatController',
        resolve: {
          chatRoom: function () {
            return 'haskell';
          }
        }
      }
    }
  }
})
.state('app.erlang', {
  url: '/erlang',
  views: {
    'erlang-view': {
      templateUrl: 'templates/app-chat.html',
      controller: 'ChatController',
      resolve: {
        chatRoom: function () {
          return 'erlang';
        }
      }
    }
  }
})
.state('app.scala', {
  url: '/scala',
  views: {
    'scala-view': {
      templateUrl: 'templates/app-chat.html',
      controller: 'ChatController',
      resolve: {
        chatRoom: function () {
          return 'scala';
        }
      }
    }
  }
});
$urlRouterProvider.otherwise('/app/node');
})
```

Note how we coupled each single tab with a given application state. In doing so, we also tell the app how it should render the view under each tab. In our case, we have a common view for each single chat, `templates/app-chat`, which is familiar to us from our previous work. Let's take a look at the following code:

```
<ion-view view-title="chat">
  <ion-content>
    <div class="list">
      <a collection-repeat="message in messages"
      class="item item-avatar"
      ng-class="{'other-chatbox' : message.external}">
        <h2>{{message.name}}</h2>
        <p>{{message.text}}</p>
      </a>
    </div>

  </ion-content>
  <div class="bar bar-footer bar-balanced">
    <label class="item-input-wrapper">
      <input id="message-input"
      type="text"
      placeholder="Message"
      ng-model="inputMessage">
    </label>
    <button class="button button-small"
    ng-click="onSend()">
      Submit
    </button>
  </div>
</ion-view>
```

Finally, add some custom CSS to the `css/style.css` file in order to adjust the formatting according to our needs; this will also be familiar, as we saw this in the previous chapter:

```
#message-input {
  width: 100%;
}

.item-avatar {
  padding-left: 16px;
}

.other-chatbox {
  text-align: right;
}
```

Your view should now look like what's shown in the following screenshot:

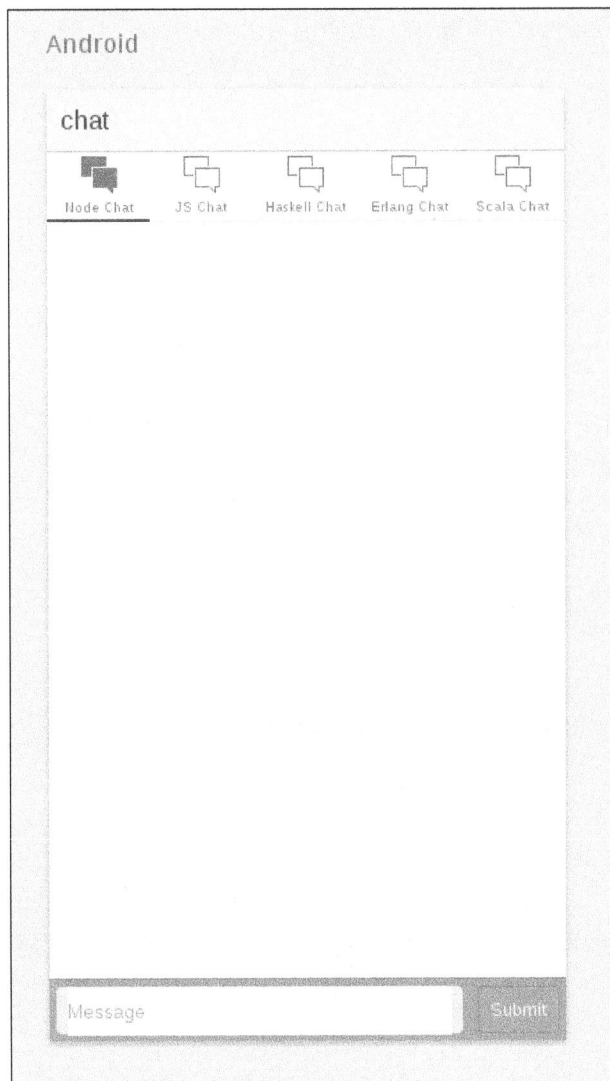

Now, let's add some actual logic to our app in order to get the actual chat logic going. We are going to implement the namespace pattern that we discussed earlier in this chapter, adding one room for each tab. First, define the following controller in the `app.controllers.js` file, as follows:

```
angular.module('ionic-chat-app-controllers', [])
.controller('ChatController', function ($scope,
ChatService, chatRoom) {
```

```
    var connection = ChatService.connect(chatRoom);
    // The chat messages
    $scope.messages = [];
    // Notify whenever a new user connects
    connection.on.userConnected(function (user) {
      $scope.messages.push({
        name: 'Chat Bot',
        text: 'A new user has connected!'
      });
      $scope.$apply();
    });
    // Whenever a new message appears, append it
    connection.on.messageReceived(function (message) {
      message.external = true;
      $scope.messages.push(message);
      $scope.$apply();
    });
    $scope.inputMessage = '';
    $scope.onSend = function () {
      $scope.messages.push({
        name: 'Me',
        text: $scope.inputMessage
      });
      // Send the message to the server
      connection.emit({
        name: 'Anonymous',
        text: $scope.inputMessage
      });
      // Clear the chatbox
      $scope.inputMessage = '';
    }
  });
```

This controller works very much like what we are used to from the previous app, with the exception that it takes as a parameter the name of the chat room that we should connect to. This name is resolved in `app.js` in conjunction with the view being resolved, as follows:

```
.state('app.scala', {
  url: '/scala',
  views: {
    'scala-view': {
      templateUrl: 'templates/app-chat.html',
      controller: 'ChatController',
      resolve: {
```

```
      chatRoom: function () {
        return 'scala';
      }
    }
  }
}
});
```

The relevant part is emphasized. We simply bind `chatRoom` to whatever the name of the corresponding language room for the view is in this case.

Finally, we need to expand the `ChatService` module in order to make sure that we can connect to an individual chat room. Open the `app.services.js` file and make sure that it has the following:

```
angular.module('ionic-chat-app-services', [])
.service('ChatService', function ChatService($rootScope) {
  function ChatConnection(chatName) {
    this.chatName = chatName;
    // Init the Websocket connection
    var socket = io.connect('http://localhost:8080/' +
    chatName);
    // Bridge events from the Websocket connection to the
    rootScope
    socket.on('UserConnectedEvent', function (user) {
      console.log('User connected:', user);
      $rootScope.$emit('UserConnectedEvent', user);
    });
    /*
    * Send a message to the server.
    * @param message
    */
    socket.on('MessageReceivedEvent', function (message) {
      console.log('Chat message received:', message);
      $rootScope.$emit('MessageReceivedEvent', message);
    });
    this.emit = function (message) {
      console.log('Sending chat message:', message);
      socket.emit('MessageSentEvent', message);
    };
    this.on = {
      userConnected: function (callback) {
        $rootScope.$on('UserConnectedEvent', function
        (event, user) {
```

```
        callback(user);
      });
    },
    messageReceived: function (callback) {
      $rootScope.$on('MessageReceivedEvent', function
      (event, message) {
        callback(message);
      });
    }
  }
}
/**
 * Establishes a new chat connection.
 *
 * @param chatName name of the chat room to connect to
 * @returns {ChatService.ChatConnection}
 */
this.connect = function (chatName) {
  return new ChatConnection(chatName);
}
});
```

In its previous incarnation, this service simply made a `socket` connection and serviced it. Here, we produce `socket` connections instead based on the namespace that we are connecting to. This allows us to set up a separate `service` instance for each individual `socket`.

That's all that we need for the client! Let's turn to the server in order to wrap things up.

Building the server

We have already seen how to create namespaces on the server. So, let's adjust our own accordingly. However, in order to make it much neater, let's do so by iterating over a list with all the names of the namespaces that we wish to create:

```
var http = require('http');
var url = require('url');
var fs = require('fs');
var server = http.createServer(function (req, res) {
  var parsedUrl = url.parse(req.url, true);
  switch (parsedUrl.pathname) {
    case '/':
```

```
    // Read the file into memory and push it to the client
    fs.readFile('index.html', function (err, content) {
      if (err) {
        res.writeHead(500);
        res.end();
      }
      else {
        res.writeHead(200, {'Content-Type': 'text/html'});
        res.end(content, 'utf-8');
      }
    });
    break;
  }
});
server.listen(8080);
server.on('listening', function () {
  console.log('Websocket server is listening on port', 8080);
});
// Connect the websocket handler to our server
var websocket = require('socket.io')(server);
// Configure the chat rooms
['node', 'javascript', 'haskell', 'erlang',
'scala'].forEach(function (chatRoom) {
  websocket.of('/' + chatRoom).on('connection',
  function (socket) {
    console.log("New user connected to", chatRoom);
    // Tell others a new user connected
    socket.broadcast.emit('UserConnectedEvent', null);
    // Bind event handler for incoming messages
    socket.on('MessageSentEvent', function (chatData) {
      console.log('Received new chat message', chatData);
      // By using the 'broadcast' connector, we will
      // send the message to everyone except the sender.
      socket.broadcast.emit('MessageReceivedEvent', chatData);
    });
  });
});
```

That's it! You can now start up your server, connect the app to server, and try it out. Pay special attention to your server console when you switch between the rooms. You will see the separate connections to separate rooms being made. Finally, see for yourself that the namespacing actually works. The messages that you send to one chat will only be visible to the users who are already connected to it.

> It is actually possible to partition the socket.io connections even further than what we did here. The socket.io connection also features the concept of rooms, which are essentially partitions of a single namespace. We recommend that you study this closely. The official documentation of socket.io contains a great deal of examples. To view this documentation, visit http://socket.io/docs/rooms-and-namespaces/.

Summary

In this chapter, you created an advanced chat application, which allows its users to chat across several rooms using the important socket.io concept of namespacing. You learned how to configure namespaces on the server itself as well as how to connect to them from the client.

In the next chapter, we will wrap up what you learned so far by looking at how we can implement a common piece of functionality — an e-commerce application.

14
Creating an E-Commerce Application Using the Ionic Framework

In this chapter, we will bring together all the knowledge that we accumulated in this book until now and implement it in an easy-to-use Ionic framework, which can be applied in our own projects.

In particular, in this chapter, we will build on the work that has already been done in *Chapter 3, Creating an API*, and we will use this work with an Ionic project, which will be accessible through an Android or iOS smartphone.

Designing our application

As part of the application development process, it is important that we understand how we will structure our application and connect it to the product API in order to achieve our final goal of creating a basic e-commerce application.

For this particular project, we will work on two main screens—the product's list items, the controller, and the product page. In addition to this, we will also focus on creating a basic functional side menu and a rudimentary purchasing option, which is nonfunctional at this stage.

Creating an Ionic project

We will start off our project by creating a project based on one of the Ionic starter templates that we didn't have an opportunity to use in the previous chapters. The side menu templates provide us with a side menu, a list item's view, and a list item's detailed view. This template should give us the necessary groundwork to create the ideal e-commerce application.

Open your terminal and input the following command:

```
$ ionic start grocerApp sidemenu
```

The preceding command will create a project folder with the sidemenu project. In order to further understand how this template works, we encourage you to first navigate to the project folder and then input the following command:

```
$ ionic serve --lab
```

This will open the browser of your choice and give you a side-by-side view of how the application will look on an Android and iOS device:

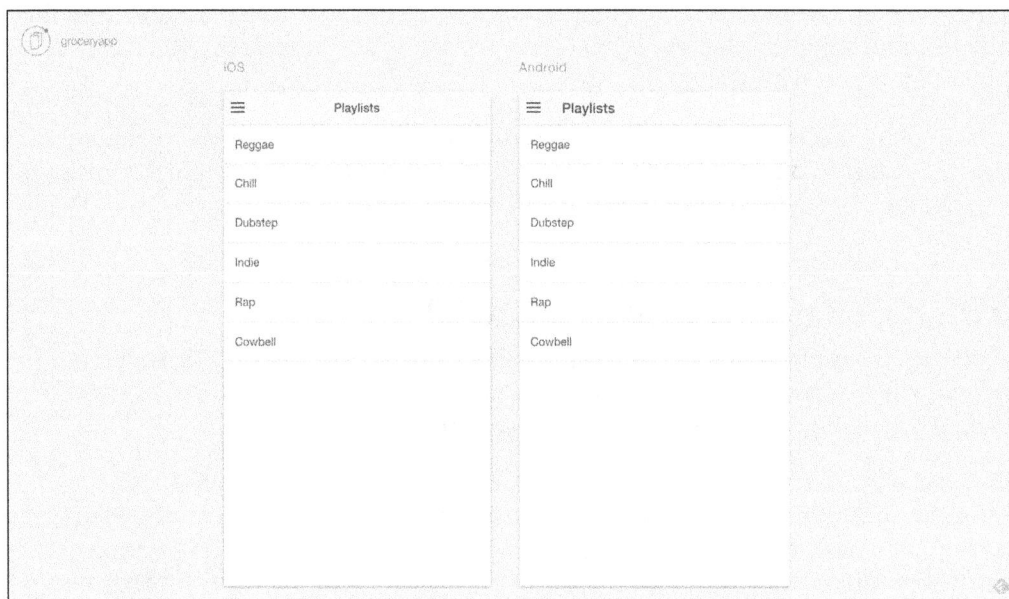

Implementing our designs

When implementing our designs, we need to reflect the necessary changes within the app's code. We will first start off by modifying our controllers.

We will first begin by navigating to `app.js`, which is available at the path `www/js/app.js`.

We will replace the existing code in `app.js` with the following:

```
angular.module('starter', ['ionic', 'starter.controllers'])

.run(function($ionicPlatform) {
  $ionicPlatform.ready(function() {
  // Hide the accessory bar by default (remove this to show the
  accessory bar above the keyboard
  // for form inputs)
  if (window.cordova && window.cordova.plugins.Keyboard) {
    cordova.plugins.Keyboard.hideKeyboardAccessoryBar(true);
    cordova.plugins.Keyboard.disableScroll(true);

  }
  if (window.StatusBar) {
    // org.apache.cordova.statusbar required
    StatusBar.styleDefault();
  }
});

.config(function($stateProvider, $urlRouterProvider) {
  $stateProvider

  .state('app', {
    url: '/app',
    abstract: true,
    templateUrl: 'templates/menu.html',
    controller: 'AppCtrl'
  })

  .state('app.search', {
    url: '/search',
    views: {
      'menuContent': {
      templateUrl: 'templates/search.html'
    }
  }
})

state('app.products', {
```

```
    url: '/products',
    views: {
      'menuContent': {
      templateUrl: 'templates/products.html',
      controller: 'ProductsCtrl'
    }
  }
})

.state('app.single', {
  url: '/products/:productId',
  views: {
    'menuContent': {
    templateUrl: 'templates/product.html',
    controller: 'ProductCtrl'
    }
  }
});
// if none of the above states are matched, use this as
the fallback
$urlRouterProvider.otherwise('/app/products');
});
```

The aforementioned code will allow us to implement the different screens that form a part of our app, namely the `products` page, the individual product, and the search functionality, which will not be implemented in the current version of the application.

The next step in our modification stage is to implement the necessary changes in our app `controllers`, which are based in the `controllers.js` file. Go ahead and replace the existing code with the following:

```
angular.module('starter.controllers', [])

.controller('AppCtrl', function($scope, $ionicModal, $timeout) {

  // Form data for the login modal
  $scope.loginData = {};

  // Create the login modal that we will use later
  $ionicModal.fromTemplateUrl('templates/login.html', {
    scope: $scope
  }).then(function(modal) {
```

```
    $scope.modal = modal;
  });

  // Triggered in the login modal to close it
  $scope.closeLogin = function() {
    $scope.modal.hide();
  };

  // Open the login modal
  $scope.login = function() {
    $scope.modal.show();
  };

  // Perform the login action when the user submits the login form
  $scope.doLogin = function() {
    console.log('Doing login', $scope.loginData);

    // Simulate a login delay. Remove this and replace with
    your login
    // code if using a login system
    $timeout(function() {
      $scope.closeLogin();
    }, 1000);
  };
})

.controller('ProductsCtrl', function($scope) {
  $scope.products = [
    { title: 'Apples', id: 1 ,price:1.00,
    image:'http://loremflickr.com/30/30/apples'},
    { title: 'Carrots', id: 2,price:2.00,
    image:'http://loremflickr.com/30/30/carrots' },
    { title: 'Tomatoes', id: 3 ,price:3.00,
    image:'http://loremflickr.com/30/30/tomatoes'},
    { title: 'Pears', id: 4, price:1.50,
    image:'http://loremflickr.com/30/30/pears' },
    { title: 'Grapes', id: 5, price:1.00,
    image:'http://loremflickr.com/30/30/grapes' },
    { title: 'Plums', id: 6, price: 2.50,
    image:'http://loremflickr.com/30/30/plums' },
    { title: 'Olives', id:7, price: 0.50,
    image:'http://loremflickr.com/30/30/olives'}
  ];
```

```
})

.controller('ProductCtrl', function($scope, $stateParams) {
});
```

As you can see in the preceding code, we declared an array of products. At this point in time, this declares a product title through the title variable, the product ID through id, and price through price. Last but not least, in order to spice things up, we also added a link to a thumbnail image generator supported by http://loremflickr.com.

The current setup will not reflect in the frontend of our mobile application because we haven't done the necessary changes in the HTML files.

We will first rename playlist.html and playlists.html to product.html and products.html respectively. We can find both of these files at the www/ templates/playlist.html and www/templates/playlists.html path.

We will then navigate to the menu.html file, which is available at the www/ templates/menu.html path.

We will replace the existing code in the preceding path with the following:

```html
<ion-side-menus enable-menu-with-back-views="false">
  <ion-side-menu-content>
    <ion-nav-bar class="bar-stable">
      <ion-nav-back-button>
      </ion-nav-back-button>

      <ion-nav-buttons side="left">
        <button class="button button-icon button-clear
        ion-navicon" menu-toggle="left">
        </button>
      </ion-nav-buttons>
    </ion-nav-bar>
    <ion-nav-view name="menuContent"></ion-nav-view>
  </ion-side-menu-content>

  <ion-side-menu side="left">
    <ion-header-bar class="bar-stable">
      <h1 class="title">Shop Menu</h1>
    </ion-header-bar>
    <ion-content>
      <ion-list>
        <ion-item menu-close href="#/app/search">
```

```
        Search
      </ion-item>
      <ion-item menu-close href="#/app/products">
        Products
      </ion-item>
      <ion-item menu-close href="#">
        Basket
      </ion-item>
    </ion-list>
  </ion-content>
  </ion-side-menu>
</ion-side-menus>
```

In the preceding code, we replaced the old reference to different template files with the more recent ones, which reflect our most recent changes.

Following this, we will proceed and modify the product.html file to to give our application a more product-like appearance. In addition to this, this page will also include an image placeholder, Product Description, Price, and a rudimentary Add to Basket button. In future iterations of the application, this will allow users to add a product to a virtual shopping basket when they wish to buy the necessary items. We will replace the existing code in product.html with the following:

```
<ion-view view-title="Product">
  <ion-content>
    <h1>Product</h1>
    <img src="http://loremflickr.com/380/160/fruits,vegetables">
    <br>
      <p>Product Description</p>
    <br>
      <p>Price</p>
    <button class="button button-balanced">
      Add to Basket
    </button>
  </ion-content>
</ion-view>
```

In the final step of modifying the HTML files, we will need to modify the products. html file to show the product title and product image using AngularJS. Replace the existing code with the following:

```
<ion-view view-title="The Grocer Shop">
  <ion-content>
    <ion-list>
      <ion-item ng-repeat="product in products">
```

```
        <a class="item item-thumbnail-left"
        href="#/app/products/{{product.id}}">
            <img src={{product.image}}>
            <h2>{{product.title}}</h2>
            <p>EUR {{product.price}} per kilogram</p>
        </a>
    </ion-item>
  </ion-list>
 </ion-content>
</ion-view>
```

In the aforementioned code, we extracted the `product.image` and `product.title` declared in `app.js` and reproduced it in the `ion-view` tag. We also personalized it and included the currency and how much the product costs per kilogram.

Setting up the product API

What we have implemented until now is a very simplified version of what we would like to achieve. Since we want to create projects that use MongoDB, Node.js and Ionic, we should take the opportunity to create an application that connects to our locally stored backend with a view of using this knowledge to connect to Internet-based servers powered by Node.js and MongoDB.

In order to take advantage of this section, you will need to follow the instructions available in *Chapter 3, Creating an API*, that are required to set up your very own Node.js server and include a basic set of data.

Once you've performed all the necessary steps, go ahead and find out what the current entries on our server are by first running the `mongodb` database. We will do this by first navigating to the `order_api` folder and running the following command:

sudo mongod

We shall start the Node.js server in our terminal by using the following command:

node api.js

If you adhered to the instructions given in *Chapter 3, Creating an API*, the following message will appear:

Up, running and ready for action!

At this point, we will open the installed REST client and pass the following command:

```
http://localhost:8080/api/products
```

If you have a response similar to the one as follows, then you should consider your attempt at creating a server to be successful:

```
[
  {
    "_id": "55be0d021259c5a6dc955289",
    "name": "Apple",
    "price": 2.5
  },
  {
    "_id": "55be0d541259c5a6dc95528a",
    "name": "Oranges",
    "price": 1.5
  },
  {
    "_id": "55be0d6c1259c5a6dc95528b",
    "name": "Pear",
    "price": 3
  },
  {
    "_id": "55be0d6c1259c5a6dc95528c",
    "name": "Orange",
    "price": 3
  }
]
```

Connecting the product API to our Ionic app

Once you have managed to get a response from the server and have the server and database up and running, you need to replace the existing array in the app.js file with the one from the local host.

Since the web server is based locally, you will need to enable cross-origin resource sharing, which is currently attainable in the easiest way through Google Chrome and by enabling the **CORS (cross-origin Resource sharing)** Chrome extension, which is available at https://goo.gl/oQNhwh. The extension is also available at the Chrome Web Store if you look for **Allow-Control-Allow-Origin: ***.

We will first start off our project by navigating to the `ionic.project` file, which is available in the `root` folder, and adding the following code to `ionic.project`:

```
"proxies": [
    {
      "path": "/api",
      "proxyUrl": "http://cors.api.com/api"
    }
  ]
```

The `http://cors.api.com/api` URL here acts as a placeholder URL in order to enable local development and cross-origin resource sharing.

This modification will help us add a proxy URL, which will allow cross-origin resource sharing.

We will also modify the `gulpfile.js` file by adding two variables and two tasks, as follows:

```
var replace = require('replace');
var replaceFiles = ['./www/js/app.js'];

gulp.task('add-proxy', function() {
  return replace({
    regex: "http://cors.api.com/api",
    replacement: "http://localhost:8080/api",
    paths: replaceFiles,
    recursive: false,
    silent: false
  });
})

gulp.task('remove-proxy', function() {
  return replace({
    regex: "http://localhost:8080/api",
    replacement: "http://cors.api.com/api",
    paths: replaceFiles,
    recursive: false,
    silent: false
  });
})
```

To make sure that the `gulpfile` functions correctly, we encourage you to make sure that `gulp` is installed correctly by running the following command:

```
sudo npm install gulp -g
```

Using `gulp`, we will also need to install `replace`. This is a `gulp` dependency, which will allow us to add the proxy functionality to the project by allowing for string replacement. This can be enabled by running the following command:

```
sudo npm install --save replace
```

In order to facilitate cross-origin resource sharing in Ionic, we will also need to use a `factory` method, which will be done by creating a new JavaScript file entitled `services` in the `www/js` folder, which contains the following code:

```javascript
angular.module('starter.services', [])
factory('Api', function($http, ApiEndpoint) {
  console.log('ApiEndpoint', ApiEndpoint);

  var getApiData = function() {
    return $http.get(ApiEndpoint.url + '/products');
  };

  return {
    getApiData: getApiData
  };
})
```

In order to create the preceding code, we will need to reference `services.js` in the `index.html` file and `app.js` file. We will add the following code to the `index.html` head tag:

```html
<script src="js/services.js"></script>
```

In addition to this, we will update the `app.js` file to include our new constant, which has already been referenced in the `services.js` file. This will be updated as follows:

```javascript
angular.module('starter', ['ionic', 'starter.controllers','starter.
services'])

.constant('ApiEndpoint', {
  url: 'http://localhost:8080/api'
})
```

In order to facilitate your coding experience, we uploaded all our code to the GitHub repository, which is available at `https://github.com/stefanbuttigieg/nodejs-ionic-mongodb/tree/master/chapter-14`.

The connection of our locally created REST API to our Angular controller will be made available to the user by updating the `product` controller entitled `ProductsCtrl`. The code needs to be updated as follows:

```
.controller('ProductsCtrl', function($scope, Api) {
  $scope.products = null;
  Api.getApiData().then(function(result) {
        $scope.products = result.data;
  });
})
```

This code modification removes the JSON array and replaces it with code that extracts data from the JSON available on the local web server and makes it available in our `controller`. The finishing touch has to be implemented through the `products.html` file. Here, we will update the file to contain a generic image placeholder. We will make slight modifications to the `products.html` file to make it work with our very own JSON file:

```html
<ion-view view-title="The Grocer Shop">
  <ion-content>
    <ion-list>
      <ion-item ng-repeat="product in products">
        <a class="item item-thumbnail-left"
        href="#/app/products/{{product.id}}">
          <img src="http://placehold.it/30x30">
          <h2>{{product.name}}</h2>
          <p>EUR {{product.price}} per kilogram</p>
        </a>
      </ion-item>
    </ion-list>
  </ion-content>
</ion-view>
```

Once the preceding code is implemented, feel free to navigate to the `root` folder of the `grocerApp` software and run the following command:

```
ionic serve --lab
```

The final app should look like this:

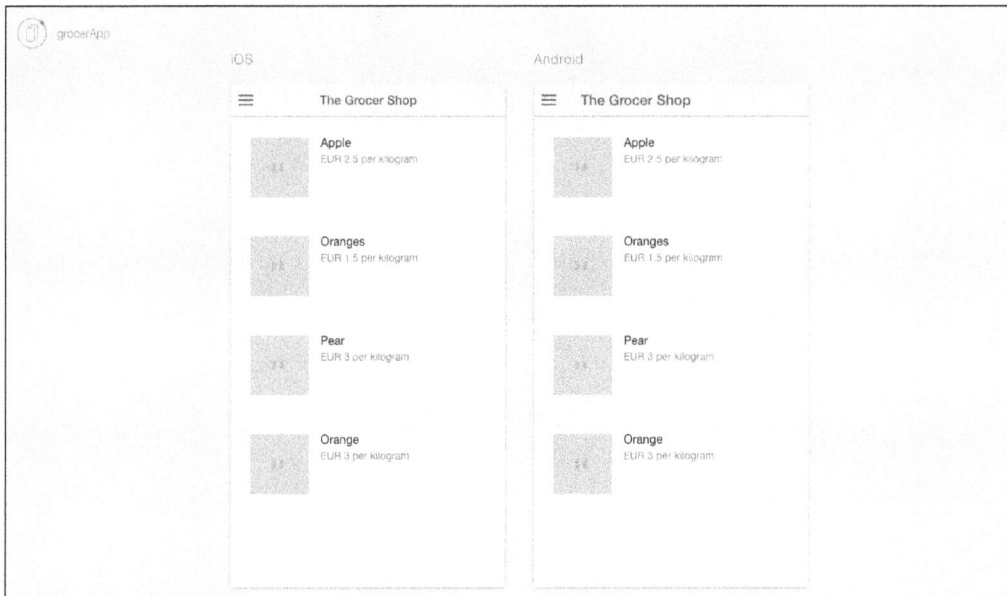

Summary

In this chapter, we brought together a number of skills that we managed to develop over the past few chapters. It's important to note that we managed to connect an API that we created from scratch and a cross-platform application that we implemented via the Ionic framework. As regards to e-commerce, there are a number of open source solutions such as **Traider.io** and **ReactionCommerce**. These solutions have expanded functionalities and are in the process of improving through community contributions, especially with regards to the REST API services. In addition to this, they make use of MongoDB and Node.js.

The Ionic framework is an ever-growing platform in all aspects. As we mentioned earlier in this book, the Ionic framework team released a number of updates with a number of new features. We are excited about this, and we believe that it's priceless to invest energy and time to further understand this platform.

Index

B

backend, Node.js 1
basic authentication service, secure app 160
  getCurrent function 162
  isAuthenticated function 162
  login function 160, 161
basic components, client-side security
  authentication tokens 152
  secure communication 153
  secure local storage 152
basic MVC project
  connection, testing 111
  controller, creating 107
  creating 104
  list view, creating 105, 106
  model, creating 112
  services 112
  services, creating 113
  view and controller, connecting 108-110
  view, creating 104, 105
bootstrap CSS framework 72

C

CentOS
  MongoDB, installing on 15
  Node.js, installing on 3
chat application
  basic app structure, setting up 172-175
  building 168-172
  ChatService function 179
  chat view, updating 181-183
  input section 176
  message view 177, 178
  requisites 168
  server, creating 168-172
  WebSockets, adding to mix 180, 181
client-side security
  about 151, 152
  basic components 152
  overview 151
collections
  adding 59, 60
  creating 25-28

Command Line Interface (CLI) 103
Cordova
  setting up, for Mac OS X 5, 6
  setting up, for Windows 6, 7
CORS (cross-origin Resource sharing)
  reference 213
CRUD (Create/Read/Update/Delete) 40
customer 60

D

database
  creating 24
Database Management System (DBMS) 22
Debian
  Node.js, installing on 3
DELETE handlers
  implementing 51
DELETE requests 43
device data
  accessing 115
  Cordova, adding to factory 116-118
  native services, accessing 115
  ngCordova 116
directives 94
Document Object Model (DOM) 129
documents
  relations, creating between 28-30

E

e-commerce application
  creating 205
  designing 205
  designs, implementing 206-211
  Ionic project, creating 206
  product API, connecting to Ionic
    app 213-216
  product API, setting up 212
Environment Variables
  setting up, on Windows 7 12
Environment Variables, for iOS
  setting up, on Mac OS X 13
expressions 94

J

Java
  installing 7, 8
Java JDK
  URL, for downloading 7
JavaScript Object Notation (JSON)
      object 106
JSON
  returning 46

L

link function
  $attr parameter 134
  $element parameter 134
  $scope parameter 134
  about 134
Linux
  MongoDB, installing on 15
  MongoDB, running on 18, 19
  Node.js, installing on 3
Loopback.js
  about 54
  URL 54

M

Mac
  Android Studio, setting up for 8-10
Mac OS X
  Cordova, setting up for 5, 6
  Environment Variables for iOS,
      setting up on 13
  Ionic framework, setting up for 5, 6
  MongoDB, running on 16, 17
  Node.js, installing on 4
Mobile/Responsive Web Design Tester 112
Model-View-Controller (MVC) pattern 92
module
  about 92, 93
  controllers 93
  services 93
MongoDB
  about 13, 22
  and Node.js, connecting 32

collections 23
connecting to 17, 24, 36
databases 23
defining 21
documents 22
in Linux 24
in Mac OS X 24
installing, on CentOS 15
installing, on Fedora 15
installing, on Linux 15
installing, on RHEL 15
installing, on Ubuntu 15
installing, on Windows 14
in Windows 24
product order database 23
querying 30
running, on Linux 18, 19
running, on Mac OS X 16
running, on Windows 18
searching, by ID 30, 31
searching, by property value 31
starting 16
starting, on Mac OS X 17
URL 14
MongoDB instance
  connecting to 17
multiroom chat application, advanced
      chat app
  basic layout, configuring 192-201
  creating 191, 192
  server, building 201, 202

N

native devices
  Android 118
  building for 118
  list view 119-121
new project
  creating 103
new tab
  adding 98
  new controller, adding 98
  state, adding 99, 100
  testing 100
  view, adding 98

Thank you for buying
Learning Node.js for Mobile Application Development

About Packt Publishing

Packt, pronounced 'packed', published its first book, *Mastering phpMyAdmin for Effective MySQL Management*, in April 2004, and subsequently continued to specialize in publishing highly focused books on specific technologies and solutions.

Our books and publications share the experiences of your fellow IT professionals in adapting and customizing today's systems, applications, and frameworks. Our solution-based books give you the knowledge and power to customize the software and technologies you're using to get the job done. Packt books are more specific and less general than the IT books you have seen in the past. Our unique business model allows us to bring you more focused information, giving you more of what you need to know, and less of what you don't.

Packt is a modern yet unique publishing company that focuses on producing quality, cutting-edge books for communities of developers, administrators, and newbies alike. For more information, please visit our website at www.packtpub.com.

About Packt Open Source

In 2010, Packt launched two new brands, Packt Open Source and Packt Enterprise, in order to continue its focus on specialization. This book is part of the Packt Open Source brand, home to books published on software built around open source licenses, and offering information to anybody from advanced developers to budding web designers. The Open Source brand also runs Packt's Open Source Royalty Scheme, by which Packt gives a royalty to each open source project about whose software a book is sold.

Writing for Packt

We welcome all inquiries from people who are interested in authoring. Book proposals should be sent to author@packtpub.com. If your book idea is still at an early stage and you would like to discuss it first before writing a formal book proposal, then please contact us; one of our commissioning editors will get in touch with you.

We're not just looking for published authors; if you have strong technical skills but no writing experience, our experienced editors can help you develop a writing career, or simply get some additional reward for your expertise.

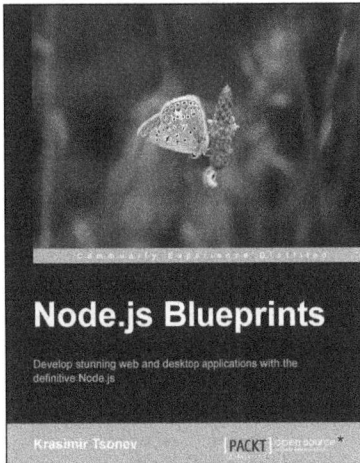

Node.js Blueprints

ISBN: 978-1-78328-733-8   Paperback: 268 pages

Develop stunning web and desktop applications with the definitive Node.js

1. Utilize libraries and frameworks to develop real-world applications using Node.js.

2. Explore Node.js compatibility with AngularJS, Socket.io, BackboneJS, EmberJS, and GruntJS.

3. Step-by-step tutorials that will help you to utilize the enormous capabilities of Node.js.

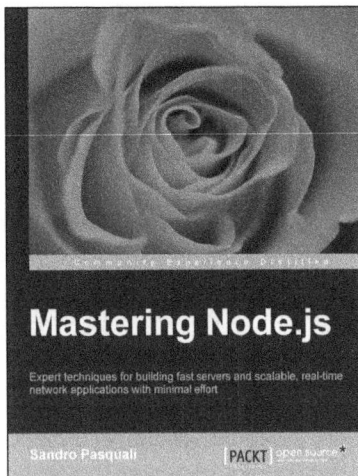

Mastering Node.js

ISBN: 978-1-78216-632-0   Paperback: 346 pages

Expert techniques for building fast servers and scalable, real-time network applications with minimal effort

1. Master the latest techniques for building real-time, big data applications, integrating Facebook, Twitter, and other network services.

2. Tame asynchronous programming, the event loop, and parallel data processing.

3. Use the Express and Path frameworks to speed up development and deliver scalable, higher quality software more quickly.

Please check **www.PacktPub.com** for information on our titles

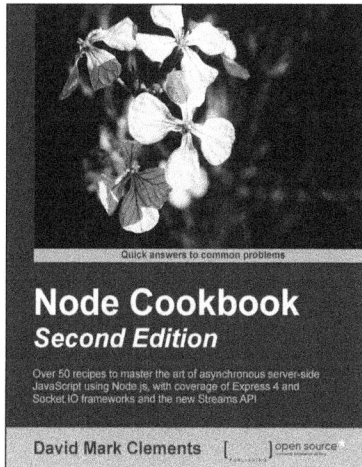

Node Cookbook

Second Edition

ISBN: 978-1-78328-043-8   Paperback: 378 pages

Over 50 recipes to master the art of asynchronous
server-side JavaScript using Node.js, with coverage
of Express 4 and Socket.IO frameworks and the
new Streams API

1. Work with JSON, XML, web sockets to make
    the most of asynchronous programming.

2. Extensive code samples covering Express 4
    and Socket.IO.

3. Learn how to process data with streams
    and create specialized streams.

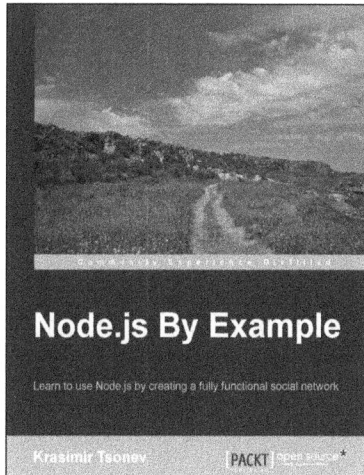

Node.js By Example

ISBN: 978-1-78439-571-1   Paperback: 220 pages

Learn to use Node.js by creating a fully functional
social network

1. Plan and implement a modern
    Node.js application.

2. Get to know the most useful
    Node.js capabilities.

3. Learn how to create complex
    Node.js applications.

Please check **www.PacktPub.com** for information on our titles

www.ingramcontent.com/pod-product-compliance
Lightning Source LLC
Chambersburg PA
CBHW061406210326
41598CB00035B/6111